THE FIREFLY GUIDE TO
MINERALS,
ROCKS
& GEMS

T0190583

RUPERT HOCHLEITNER
FIREFLY BOOKS

A FIREFLY BOOK

Published by Firefly Books Ltd., 2019
© 2019 Firefly Books Ltd.
© 2017 Franchkh-Kosmos Verlags-GmbH & Co. KG, Stuttgart, Germany
Original title: Hochleitner, Welcher Stein ist das?

2nd printing, 2023

Library of Congress Control Number: 2019938807

Library and Archives Canada Cataloguing in Publication
Title: The Firefly guide to minerals, rocks & gems / Rupert Hochleitner.
Other titles: Welcher stein ist das? English | Firefly guide to minerals, rocks and gems
Names: Hochleitner, Rupert, author.
Description: Translation of: Welcher stein ist das?. | Includes index.
Identifiers: Canadiana 20190095423 | ISBN 9780228102281 (softcover)
Subjects: LCSH: Minerals—Identification. | LCSH: Rocks—Identification. | LCSH: Gems—Identification.
| LCSH: Mineralogy—Handbooks, manuals, etc. | LCSH: Petrology—Handbooks, manuals, etc. | LCGFT:
Field guides. | LCGFT: Handbooks and manuals.
Classification: LCC QE365 .H6313 2019 | DDC 549—dc23

Published in Canada by
Firefly Books Ltd.
50 Staples Avenue, Unit 1
Richmond Hill, Ontario
L4B 0A7

Published in the United States by
Firefly Books (U.S.) Inc.
P.O. Box 1338, Ellicott Station
Buffalo, New York
14205

Translation: Travod International Ltd.

Printed in China | E

The Firefly Guide to Minerals, Rocks and Gems

How to use this book

For each rock and mineral described, you'll find details about where they can be found, as well as important information about their properties.

The main image shows a general view of a typical specimen of the particular rock or mineral type described. The characteristics important for identification are highlighted in the key. The other pictures show noteworthy details and further variations. In the margin, you can see the most important features that help with the identification of the rock or mineral.

Habits
This photo shows other habits, or external shapes, of the mineral.

Similar types
This box contains information on how to distinguish between minerals that are easily confused.

Occurrenc thermal v oxidation iron-man posits, as deposits in phic rocks.

Identifying characteristics
Important properties of the rocks and minerals can be found in the margin to help with identification.

> **Hardness** 3 1
> **Density** 3.3–3
> **Luster** Vitreou
> **Cleavage** Perf bohedral
> **Fracture** Unev
> **Tenacity** Britt

Crystal drawing
Shows a typical and frequently occurring crystal form of the mineral. It also indicates in which crystal system the mineral is classified.

Trigonal crys

140

Color code
In the identification section, minerals are sorted into 7 color groups according to their streak color and ordered according to increasing hardness. Rocks are listed in a separate chapter (see also p. 1).

Limonite

Rhodochrosite, Raspberry Spar

$MnCO_3$

Rhodochrosite is pink to red; in rare cases, it is yellowish. It forms rhombohedral or scalenohedral habits, which are often rounded ("shaped like ice grains"). Often you can also find spherical, reniform and radial aggregates, it is also stalactic, crusty, coarse.

Similar minerals
In contrast to rhodochrosite, calcite reacts with diluted cold hydrochloric acid; sometimes it is not possible to distinguish rhodochrosite from manganese-containing dolomite, which can also pink, by simple means. Beautifully colored, dense rhodochrosite is used to make jewelry.

Intergrowth of curved rhombohedron

Scalenohedral

Rounded surfaces

Name
The international name of the mineral is given. In some cases, where applicable, an alternate name is provided.

Chemical composition
The chemical composition shows which elements make up the mineral.

Interesting facts
The text describes in which habits the mineral can occur. There are other interesting facts too, such as information about how the mineral is used.

Additional photo
These small photos show other details or habits.

Typical crystal form
Here you will find information about the specific crystal form.

What rock is this?

This question arises time and again, whether you pick up a pebble during a walk, find a crystal in the mountains, discover golden or shining silver fragments in the waste rock piles of an ore mine, stumble over a curbstone, or when you look at a beautiful piece of jewelry. Inevitably, you will want to know: What mineral is this? What rock do I have in front of me? What is this colorful gemstone that sparkles so beautifully?

This book will help to answer these questions — as your constant companion on hikes or trips, while mountaineering or collecting minerals, in quarries and at waste rock piles, at mineral shows and also at jewelers.

With the exception of native mercury, minerals are always solid. No matter how good mineral water tastes, no matter how many minerals the label indicates, it is liquid and, therefore, not a mineral. Anything man-made, from window glass to quartz crystal in watches to artificial diamonds, is not a mineral.

A mineral must always be a naturally occurring substance.

The concept of a crystal is somewhat different. Crystals are solid chemical substances whose atoms are arranged in a uniform and ordered structure. This ordered arrangement of atoms is reflected in the flat, regular surfaces which form the boundaries of a crystal. Almost all minerals are crystals, even if it is sometimes not externally apparent. There are only a few minerals whose atoms are not structurally arranged in the form of a crystal lattice — these are called amorphous. The best-known example is opal, which, in contrast to quartz (which is composed almost in the same way), cannot form crystals.

Pyromorphite crystals with curved prism surfaces are known as "Emser Barrels."

Gemstones are minerals that are cut for jewelry purposes. In order to be considered a gemstone, a mineral must meet various requirements, but primarily, it must be beautiful, i.e. satisfy aesthetic requirements. This means that it should be beautifully colored and should shine and sparkle as much as possible when polished. The latter is even more important if the mineral, such as a diamond, is normally colorless.

Agates that resemble landscapes after being polished are called landscape agates.

Rocks can be described as large geological formations that are composed of individual fragments of one or more different mineral types. Marble, for example, is made up of many grains of a single mineral: calcite. Granite, on the other hand, is composed of three mineral types: feldspar, quartz and mica.

Porphyry granite with large potassium feldspar crystals.

Mineral properties

In order to identify minerals, it is necessary to recognize their properties. Each type of mineral has a number of properties, all of which are unique in their combination for the respective mineral. This means that in order to reliably identify a mineral, it is necessary to examine as many of its properties as possible. For some properties, such as hardness or streak color, this is easy and no tools are required or the tools are readily available. In the case of other properties, such as chemical composition, an exact identification requires a considerable amount of equipment to which the average individual doesn't typically have access.

For this reason, this book highlights the properties that are as easy as possible to determine and that can normally lead to the reliable identification of a mineral.

Streak color

The streak color is obtained by drawing a line with the mineral by scraping it on an unglazed and somewhat rough ceramic plate. The color of the resulting mark is characteristic to the type of mineral. Different specimens of the same type of mineral may differ in color and habit, but their streak color is always the same. Fluorite comes in a wide range of colors such as colorless, yellow, green, blue, brown, pink or purple, yet its streak color is always white.

Streak plates can be inexpensively purchased at mineral or crystal stores (usually for about $2).

Hardness

All minerals can be classified according to their hardness. The hardness of a mineral is determined by its resistance to being scratched by another mineral. Since this property is specific to each mineral, it is used in this book alongside the streak color as the most important characteristic of identification and organization.

The simplest way to determine the hardness of a mineral you are trying to identify is to compare it with the minerals on the Mohs' scale. This scale consists of a sequence of ten minerals, each of which scratches all minerals preceding it.

Mohs' scale

1	Talc	*Can be scratched with a fingernail*	
2	Gypsum		*Can be scratched with a knife*
3	Calcite		
4	Fluorite		
5	Apatite		
6	Feldspar		
7	Quartz		
8	Topaz	*Scratches glass*	
9	Corundum		
10	Diamond		

Hardness scales, i.e. a compilation of the nine test minerals (diamond is not required as it's the hardest substance), can be purchased in the mineral trade.

To determine the hardness, proceed as follows:

First, take a mineral of medium hardness, such as apatite (hardness = 5). Then examine whether the mineral to be identified can be scratched by it. If it can, proceed to the next softer test mineral until you reach one that can no longer scratch the mineral to be identified. Conversely, if you can't scratch the test mineral using the mineral to be identified, then both have the same hardness. You've reached your target. If, on the other hand, you can't scratch the mineral to be identified using the first chosen test mineral of medium hardness, then simply proceed to the next harder one. In this way, you can determine the hardness of each mineral according to the Mohs' scale.

Always check the hardness using sharp edges and on a freshly broken surface. Always wipe away any dust after scratching to make sure that it has in fact been scratched and that it's not simply the test mineral that has rubbed off.

Important: Always cross check the hardness tester! If the test mineral scratches the mineral to be identified, you must always check to see if the test mineral has been scratched as well. It's the only way you can be sure.

Tenacity

Tenacity describes how a mineral reacts when it is scratched or bent. Most minerals are brittle. When a brittle mineral is scratched (using a steel needle, for example), it results in a fine powder. When hammered, a brittle mineral can also result in a fine powder as well as small crumbs (e.g. galena). If it is possible to create a scratch mark without producing a powder, for example like cutting butter with a knife, the mineral is called sectile (e.g. argentite, gold). In addition, gold can be pounded into sheets. Such minerals are called malleable.

Other minerals, such as mica, are elastic. This means the mineral is capable of being bent and will return to its original shape once pressure is released. Flexible minerals, such as gypsum, on the other hand, remain in their new position after bending.

Gypsum crystals can be bent carefully but will not return to their original position after bending.

Color

At first glance, color seems to be the most useful property of a mineral. Unfortunately, that's not the case. There are some minerals that are very distinctive by their color, such as green malachite or blue azurite. However, a large percentage of the minerals can be found in a wide variety of colors. Quartz can be colorless, pink, purple, brown, black or yellow and diamonds can be white, yellow, green, brown, blue and black. In addition, some minerals form a different surface color when exposed to air. For example, a freshly broken piece of bornite has a metallic pink appearance. But after a few hours, its surface oxidizes, giving it a shimmering blue-red-green appearance. The color of a mineral must, therefore, always be tested on a fresh spot.

Luster

Each unprocessed mineral has a specific luster that is characteristic of the mineral type in question. However, this luster is difficult to measure. It can only be described in comparison to everyday objects.

Vitreous luster resembles the luster of simple window glass. This is the most common one.

Metallic luster is similar to that of polished metal, such as aluminum foil.

Silky luster refers to a luster that is comparable to the undulating shimmer of light on natural silk.

Pitchy luster has a tar-like appearance, comparable to the lumps of tar that can be seen during road repair work.

Greasy luster resembles the luster of greasy stains on paper.

Adamantine luster refers to the brilliant luster found in both cut diamonds and lead (crystal) glass.

Pearly luster refers to a luster similar to the inside of a mollusk shell (mother-of-pearl), which exhibits a whitish luster with a colored sheen of light.

Cubic cleavage of halite.

Density

The density, or specific gravity, is the weight of a mineral per unit volume (expressed in grams per cubic centimeter). Measuring density is not easy and requires precise equipment. Nevertheless, density can be used as an identifying characteristic. Simply holding it to feel the weight can determine whether a mineral is light (density below 2), normal (density around 2.5), heavy (density above 3.5) or very heavy (6 or higher). An even better way to estimate the density is to take an equal piece of another mineral with a known density in the other hand and compare them.

Cleavage and fracture

When a mineral is shattered (e.g. with a hammer) or broken, different looking fracture surfaces are created depending on the type of mineral. The mineral can have flat, smooth cleavage planes or can be broken into identical geometric bodies. Galena, for example, splits into small cubes and calcite into small rhombohedra. In some cases, the angles at which cleavage planes intersect each other are important in identifying a mineral. Augit, for example, can be easily distinguished from hornblende, a similar mineral, because its cleavage planes intersect at an angle of about 90°. Hornblende, on the other hand, has a cleavage angle of about 120°.

Cleavage can show different qualities from "perfect" to "indiscernible." The latter means that a cleavage does exist, but it is normally not recognizable by simple means.

The keyword **fracture** is used to describe any separation plane other than a cleavage plane. Depending on the surface appearance, the fracture can be described as conchoidal (e.g. quartz or obsidian), splintery (e.g. calcite or feldspar), uneven (e.g. feldspar) or hackly (e.g. gold or silver).

The rock glass obsidian perfectly illustrates a conchoidal fracture.

Fluorescence and phosphorescence

Exposing some minerals to ultraviolet light can cause them to glow more or less strongly in various colors. Some minerals will continue to glow for a few seconds after the UV source is switched off. This phenomenon is called phosphorescence. Both properties are generally not characteristic properties of a mineral. Individual samples of the same mineral type can show completely different fluorescent colors, while some samples may not even fluoresce at all.

Be careful when handling UV light. UV light (especially short-wavelength) can damage the eyes. Always wear safety glasses!

Origin and occurrence of rocks and minerals

Minerals take many thousands to hundreds of thousands of years to grow.

The formation of minerals is divided into three different formation cycles:

The igneous cycle comprises rocks and minerals that are formed from molten magma either inside the earth (plutonic rocks) or on the earth's surface (volcanic rocks). **Plutonic rocks** are characterized by the fact that they are relatively coarse-grained, which means that the individual grains within the matrix can be seen with the naked eye.

Volcanic rocks are very fine-grained and individual grains within the matrix cannot be seen with the naked eye or with a magnifying glass.

In the sedimentary cycle, minerals are primarily formed when minerals or rocks are weathered, transported by water or wind and later deposited again.

Sedimentary rocks are often clearly stratified, individual crystals of the rock components are not discernible. Unlike the other rocks, sedimentary rocks often contain fossils.

In the metamorphic cycle, rocks and minerals are formed through changing pressure and temperature conditions at a certain depth below the earth's surface.

Metamorphic rocks are often clearly layered and folded, individual crystals of the rock components are usually discernible.

Granite composed of feldspar, quartz and mica (here black biotite).

Igneous formations

Intramagmatic deposits are mineral accumulations within plutonic rock bodies. The metals chromium, platinum and nickel in particular are extracted from such deposits. Kimberlite pipes represent a special type of mineral occurrence in igneous rocks. Huge volcanic vents are filled with kimberlite, a special rock that often contains embedded diamond crystals.

Pegmatite is a very coarse-grained rock that has filled gaps in an older rock body. It is primarily composed of feldspar, quartz and mica. Feldspar is extracted as a raw material for the porcelain industry, while mica is used as an insulating material and, more recently, for the production of car paints.

In addition, pegmatite often contains a wide range of minerals, including gemstone minerals, which are embedded as large crystals in the rock. These minerals include beryl, topaz, tourmaline and many more.

Pneumatolytic deposits are formed from hot gases deep within the earth. Minerals that can occur in such formations include cassiterite (tin-ore), fluorite, topaz and tourmaline. Pneumatolytic deposits are primarily used to extract tin, and tungsten to a lesser extent.

Hydrothermal veins

Veins fill voids in rocks with minerals and are younger than the rock. These veins often contain open cavities in which crystals can grow freely, including gemstone minerals such as amethyst. Hydrothermal veins contain important ore minerals from which metals are extracted, such as copper, zinc, lead, silver or gold.

Alpine-type fissures present a special case: These cracks and fissures within rocks contain beautiful and sometimes very large specimens of rock crystal, smoky quartz, citrine, hematite or feldspar.

Alternating strata of ore minerals (here sphalerite) and gangue (here quartz) are typical for hydrothermal veins.

Volcanic formations

During the cooling and solidification process of molten lava, gases contained within the melt are released. A portion escapes from the surface of the lava flow, while a portion remains stuck in the rapidly solidifying rock in the form of "gas bubbles." These bubbles form more or less round cavities that can be many centimeters, rarely even meters in size. While the already solid rock continues to cool, hot solutions can seep through and fill these cavities with mineral formations. Huge deposits of such mineral formations in Brazil and Uruguay provide large quantities of amethyst and agate. Many zeolite minerals, such as phillipsite, chabasite or stilbite, can also be found in these cavities.

An amethyst druse is a cavity within volcanic rock that is filled with amethyst crystals.

Sedimentary formations
Oxidation / cementation zone

Places where deposits reach the earth's surface are subject to major changes in appearance and mineral content. The vein no longer contains sulfidic ores; the most common mineral is limonite, a ferric hydroxide, with which minerals formed through oxidation such as malachite, azurite, wulfenite, vanadinite, zinc spar and many others can be found intergrown or growing in its cavities.

Placer

Placer deposits are accumulations of minerals exposed through the weathering of rocks or deposits that are transported by water, concentrated and deposited again. They are primarily minerals characterized by their high specific weight and chemical resistance, such as gold, platinum, garnet, ilmenite, rutile, monazite and numerous gem minerals such as diamond, ruby, sapphire, chrysoberyl, topaz, spinel and many others.

Metamorphic formations

Typical minerals found in metamorphic rocks, especially marbles, are ruby and spinel and, to a lesser extent, sapphire. Gneisses or mica schists sometimes contain beautiful garnet or even emerald crystals.

Collecting rocks and minerals

The easiest, albeit most expensive, way to create a mineral collection is to buy the minerals. For this purpose, there are special shops located in cities, as well as in tourist areas. You can also find an impressive selection at mineral shows.

As a general rule, however, it is cheaper, and more satisfying but also more strenuous, to collect your minerals yourself. In fact, this is really the only way to create a rock collection, since rocks for collection purposes are rarely available for sale.

Tools such as a geologist's hammer, a club hammer for smashing larger pieces and various chisels

Rocks are collected as so-called hand specimens. The piece pictured is a diorite.

are needed to collect them. Some of these tools can be purchased at a hardware store. Tools specially made for collecting minerals, such as hardness scales, streak plates and magnifying glasses, can be found in specialty shops.

Identifying minerals

1. Check the streak color to determine which section of the book you need to search in.
2. Determine the hardness. This minimizes the number of potential minerals within the streak color group.
3. Check the other properties specified in the text. Under "Similar minerals" you will find other minerals that could be confused with your favorite mineral and which characteristics can help you to distinguish between them.

Identifying rocks

First of all, determine which group of rocks you are dealing with. Use the descriptions on p. 12 to do this.

If you know which group the rock belongs to, try to determine the constituent mineral(s). Diluted hydrochloric acid is often useful in differentiating between limestone and other similar looking rocks.

Once you have identified the constituent mineral(s), you can assign the rock to one of the rock types shown. Notable secondary minerals sometimes allow a more precise determination (e.g. hornblende granite).

Chalcanthite, Copper Vitriol
$CuSO_4 \cdot 5\,H_2O$

Occurrences In the oxidation zone of sulfidic copper deposits.

> **Hardness** 2.5
> **Density** 2.2–2.3
> **Luster** Vitreous
> **Cleavage** Hardly recognizable
> **Fracture** Conchoidal
> **Tenacity** Brittle

Blue chalcanthite sometimes forms prismatic to lenticular crystals. More often it forms stalactitic aggregates or crusts and coarse masses. It is formed during the weathering of copper-bearing ores; its formation depends on precipitation. Larger quantities are only found in dry places, where it seldom rains. Chalcanthite is easily soluble in water; the solution is characteristically blue. Dissolved copper vitriol is still used in some countries today, as a spraying agent in winegrowing.

Similar minerals
In contrast to chalcanthite, azurite is not water-soluble.

Triclinic crystal form

spherical aggregates

curly aggregate

Liroconite, Lentil-Ore
$Cu_2Al(AsO_4)(OH)_4 \cdot 4\,H_2O$

Occurrences In the oxidation zone of copper deposits.

> **Hardness** 2–2.5
> **Density** 2.95
> **Luster** Vitreous
> **Cleavage** Poor
> **Fracture** Conchoidal
> **Tenacity** Brittle

Liroconite forms blue to blue-green tabular, typically lenticular, crystals and coarse coats, often in limonite druses. It is formed during the weathering of copper ores containing arsenic. In some sources it is referred to as "lentil-ore" due to its lenticular crystals.

Monoclinic crystal form

Similar minerals
Azurite and malachite have a different color and effervesce with hydrochloric acid, the lenticular crystal form is very characteristic.

lenticular crystals

tabular crystals

Linarite

$Pb Cu[(OH)_2/SO_4]$

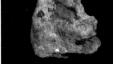

Linarite forms ink blue prismatic to less often tabular crystals with numerous faces and is frequently crusty, earthy. It usually occurs in the company of cerussite and is formed where lead and copper ores weather together. When dabbed with hydrochloric acid, linarite turns white on the surface due to the formation of bleach chloride.

Occurrences In the oxidation zone of lead deposits.

> **Hardness** 2.5
> **Density** 5.3–5.5
> **Luster** Vitreous
> **Cleavage** Recognizable
> **Fracture** Conchoidal
> **Tenacity** Brittle

columnar crystals

Similar minerals While linarite turns white, azurite effervesces when dabbed with hydrochloric acid.

prismatic crystal

Monoclinic crystal form

Boleite

$Pb_9Cu_8Ag_3Cl_{21}(OH)_{16} \cdot H_2O$

boleite-like cumengeite crystals

The intense blue crystals resemble cubes. They can be raised or embedded or they can form regular crusts. Particularly well-formed crystals are found within cavities of ancient slags.

Occurrences In the oxidation zone of copper deposits.

> **Hardness** 3–3.5
> **Density** 5.1
> **Luster** Vitreous
> **Cleavage** Perfect
> **Fracture** Conchoidal
> **Tenacity** Brittle

Similar minerals Diaboleite and cumengeite are very similar and often occur together with boleite. However, they have different crystal forms.

cubes

Tetragonal crystal form, pseudocubic

Azurite
$Cu_3[OH/CO_3]_2$

tabular crystal

malachite

Occurrences *In the oxidation zone of copper deposits.*

> **Hardness** *3.5*
> **Density** *3.7–3.9*
> **Luster** *Vitreous*
> **Cleavage** *Perfect*
> **Fracture** *Conchoidal*
> **Tenacity** *Brittle*

Azurite forms columnar to tabular crystals, in spherical groups and crusts or radial aggregates, but most frequently it is coarse and earthy. Because of its blue color, the mineral was also used as a color pigment in painting. It was far less valuable than lapis lazuli because of its lower durability. Through contact with atmospheric oxygen, blue azurite can transform into green malachite while retaining its crystal form. In paintings, for example, this means that blue skies turn green over time.

Monoclinic crystal form

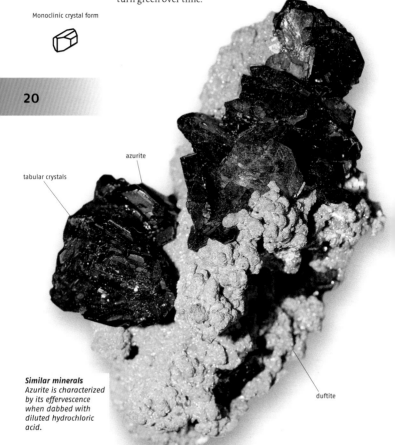

azurite

tabular crystals

duftite

Similar minerals
Azurite is characterized by its effervescence when dabbed with diluted hydrochloric acid.

Connellite

$Cu_{19}Cl_4SO_4(OH)_{32} \cdot H_2O$

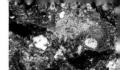

Connellite forms blue acicular crystals, often growing in tufts. In contrast to connellite, buttgenbachite, which is indistinguishable by simple means, contains nitrogen. Sometimes connellite can also be found in ancient slags.

spherical tufts of needles

Occurrences In the oxidation zone of copper deposits.

> **Hardness** 3
> **Density** 3.41
> **Luster** Vitreous
> **Cleavage** None recognizable
> **Fracture** Conchoidal
> **Tenacity** Brittle

radial aggregate

fibrous crusts

buttgenbachite

Hexagonal crystal form

Similar minerals
Azurite is never so acicular; cyanotrichite is hardly distinguishable by simple means, but it is a little brighter blue.

Cyanotrichite, Velvet Copper Ore

$Cu_4Al_2[(OH)_{12}/SO_4] \cdot 2 H_2O$

The intense blue cyanotrichite forms acicular to long tabular, capillary, tufted, radial habits. It often forms velvety coatings on the parent rock, which is why it is also called velvet copper ore.

Similar minerals
Azurite is much darker; connellite cannot be distinguished by simple means; agardite is somewhat greenish and doesn't have a blue streak.

Occurrences In the oxidation zone of copper deposits.

> **Hardness** 3.5–4
> **Density** 3.7–3.9
> **Luster** Vitreous to silky
> **Cleavage** None
> **Fracture** Uneven
> **Tenacity** Brittle

velvety crust

spherical tufts of needles

Orthorhombic crystal form

limonite

Langite
$Cu_4(SO_4)(OH)_6 \cdot 2 H_2O$

Occurrences In the oxidation zone of copper deposits and in old slags, mostly as a very young formation.

> **Hardness** 3–4
> **Density** 3.48–3.5
> **Luster** Vitreous
> **Cleavage** Poor
> **Fracture** Conchoidal
> **Tenacity** Brittle

Orthorhombic crystal form

Langite forms blue to greenish blue crystals, crystal druzy, dendritic aggregates and especially crusty coatings. It is often a very young formation and can be found on tunnel walls and sometimes even on wood used to build the tunnels.

tabular crystals

reniform crust

Similar minerals
Azurite effervesces when dabbed with hydrochloric acid and is darker blue; linarite is darker blue and turns white when dabbed with hydrochloric acid.

Cornetite
$Cu_3PO_4(OH)_3$

Occurrences In the oxidation zone of copper deposits.

> **Hardness** 4.5
> **Density** 4.1
> **Luster** Vitreous
> **Cleavage** None
> **Fracture** Uneven
> **Tenacity** Brittle

Orthorhombic crystal form

The blue crystals of cornetite are short prisms, often rounded. It forms crusts and typically radial, or sun-shaped, aggregates on the source rock. Less often are formations in the form of stalactites.

sun-shaped aggregate

reniform crust

Similar minerals
Azurite effervesces when dabbed with hydrochloric acid; linarite occurs in a different paragenesis and turns white when dabbed with hydrochloric acid.

Lazurite, Lapis Lazuli

$Na_8[S/(AlSiO_4)_6]$

Lapis lazuli rarely forms embedded rhombic dodecahedrons, mostly coarse, granular, dense masses. It is characterized by its blue color in combination with white calcite and is generally speckled with golden pyrite inclusions. The aesthetic appearance makes lapis lazuli a popular gemstone that has even been found in the tombs of Egyptian pharaohs. Because of its lightfastness, lapis lazuli is the most valuable natural blue pigment and has been used especially in the depictions of the Mother of God.

Occurrences In sodium-rich marbles.

> **Hardness** 5–6
> **Density** 2.38–2.42
> **Luster** Vitreous, while its fracture is greasy
> **Cleavage** Hardly recognizable
> **Fracture** Conchoidal
> **Tenacity** Brittle

cabochon

pyrite

Similar minerals Azurite effervesces when dabbed with diluted hydrochloric acid. Colored jasper, also called "German lapis," has no pyrite inclusions.

Isometric crystal form

cubic crystal

lazurite

calcite

Crossite
$Na_2(Mg,Fe)_3(Fe,Al)_2[(OH)_2|Si_8O_{22}]$

Occurrences In sodium-rich crystalline schists.

> **Hardness** 6
> **Density** 3.1–3.2
> **Luster** Adamantine
> **Cleavage** Perfect
> **Fracture** Conchoidal
> **Tenacity** Brittle

The blue-gray crossite is a typical mineral of the blueschist facies, formed from rocks that were quickly transported to great depths. There it forms prismatic, tabular crystals, fibrous, acicular, often radial masses. Crossite-rich blueschists are sometimes used as valuable decorative stones.

Similar minerals Pumpellyite and epidote are green; glaucophane is indistinguishable by simple means.

radial aggregate

crystal fibers

Monoclinic crystal form

acicular crystals

Glaucophane
$Na_2Mg_3Al_2[(OH)_2|Si_8O_{22}]$

Occurrences In sodium-rich crystalline schists.

> **Hardness** 6
> **Density** 3–3.1
> **Luster** Vitreous
> **Cleavage** Perfect
> **Fracture** Conchoidal
> **Tenacity** Brittle

Dark-blue glaucophane is the most common species of the sodium amphibole group. It is a typical mineral of the blueschist facies, formed from rocks that were quickly transported to great depths. There it forms prismatic, tabular crystals, fibrous, acicular, often radial masses. Blue glaucophane that is combined with green fuchsite is particularly attractive.

radial aggregate

Monoclinic crystal form

acicular crystals

Similar minerals Pumpellyite and epidote are green; crossite is indistinguishable by simple means.

Kermesite, Red Antimony

Sb_2S_2O

Kermesite forms metallic red acicular, rarely prismatic crystals, which often grow together to form bundles of needle-like or radial aggregates. A typical accompanying mineral is metallic gray antimonite.

Occurrences *In the oxidation zone of antimony deposits.*

> ***Hardness*** 1–1.5
> ***Density*** 4.68
> ***Luster*** Vitreous to adamantine
> ***Cleavage*** Hardly recognizable
> ***Fracture*** Fibrous
> ***Tenacity*** Brittle

adamantine luster

acicular crystals

radial bundles

capillary

quartz

Similar minerals
In the past, all needle-like shiny ore minerals were called antimony. Kermesite is unmistakable due to its red color when paragenesis with antimonite is observed.

Monoclinic crystal form

Hutchinsonite

$(Tl,Pb)_2(Cu,Ag)As_5S_{10}$

Hutchinsonite forms prismatic to acicular crystals and, more rarely, fibrous and radial aggregates. Larger crystals are translucent blackish and red, while smaller crystals and coatings are cherry red.

Occurrences *In hydrothermal copper-silver deposits with high arsenic contents.*

Similar minerals
Enargite and realgar don't have a red streak; miargyrite, proustite and pyrargyrite are harder.

orpiment

prismatic crystals

> ***Hardness*** 1.5–2
> ***Density*** 4.6
> ***Luster*** Adamantine
> ***Cleavage*** Poor
> ***Fracture*** Conchoidal
> ***Tenacity*** Brittle

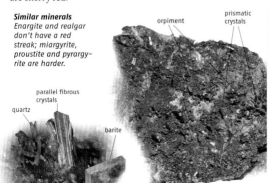

parallel fibrous crystals

quartz

barite

Orthorhombic crystal form

Erythrite, Cobalt Bloom

$Co_3(AsO_4)_2 \cdot 8\,H_2O$

Occurrences In the oxidation zone of cobalt-rich deposits.

- > ***Hardness*** 2
- > ***Density*** 3.07
- > ***Luster*** Vitreous, pearly on cleavage planes
- > ***Cleavage*** Perfect
- > ***Fracture*** Uneven
- > ***Tenacity*** Brittle, thin flexible sheets

Erythrite forms tabular to acicular crystals and spherical bundles. More frequently, however, the mineral is earthy and coarse. It is characterized by its intense purple-red to pink color. Such coatings of erythrite are always a clear indication of cobalt-containing ores. A frequent accompanying mineral, in addition to various cobalt ores, is native bismuth. Cobalt ores, together with cobalt bloom, were roasted and melted with quartz to form the so-called smalt, which, when finely ground, produced the cobalt blue pigment. Because of its resistance to heat, it was used particularly in porcelain painting.

Monoclinic crystal form

Similar minerals
Erythrite's characteristic purple-red to pink color doesn't allow for confusion; roselite and wendwilsonite have a completely different crystal form.

coating of cobalt bloom

quartz crystal

tabular crystals

adjoining rock

Miargyrite

AgSbS₂

Miargyrite forms thick tabular to
blocky crystals but is more often
simply coarse. Its color is gray to
black, often with clearly
recognizable red internal
reflections.

quartz

short prismatic
crystal

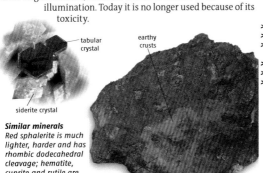

tabular crystals

Similar minerals
*Stephanite has a
gray to black streak;
proustite is intense
red; pyrargyrite is
slightly reddish;
hutchinsonite is
much softer.*

Occurrences *In hydrothermal silver ore
veins, especially in the
cementation zone.*

> **Hardness** 2.5
> **Density** 5.25
> **Luster** *Metallic*
> **Cleavage** *None recog-
> nizable*
> **Fracture** *Conchoidal*
> **Tenacity** *Brittle*

Monoclinic crystal form

Cinnabar, Cinnabarite

HgS

Cinnabar is the most important mercury ore. It seldom forms
intense red crystals, but mostly coarse to granular cherry-red to
brown-red colored masses. Caution! Cinnabar is poisonous. Due
to its bright red color, cinnabar was once used in book
illumination. Today it is no longer used because of its
toxicity.

tabular
crystal

earthy
crusts

siderite crystal

Similar minerals
*Red sphalerite is much
lighter, harder and has
rhombic dodecahedral
cleavage; hematite,
cuprite and rutile are
harder.*

dull

Occurrences *In
low-temperature
hydrothermal veins, at
outlets of volcanic gases
on surrounding rocks.*

> **Hardness** 2–2.5
> **Density** 8.01
> **Luster** *Adamantine, fine-
> grained often matte*
> **Cleavage** *Perfect*
> **Fracture** *Splintery*
> **Tenacity** *Brittle*

Trigonal crystal system

Proustite, Light Red Silver Ore
Ag_3AsS_3

calcite crystals

scalenohedral crystal

Occurrences In hydro-thermal silver ore veins.

> **Hardness** 2.5
> **Density** 5.5–5.7
> **Luster** Adamantine to metallic
> **Cleavage** Poorly recognizable
> **Fracture** Conchoidal
> **Tenacity** Brittle

Proustite forms intensively grown red prismatic to pyramidal crystals. When exposed to light for a long time, they become darker to almost black. Together with other silver sulfides, proustite was an important mineral found in rich ore deposits mined in ancient times. Proustite is relatively rare and no longer considered important, however, it is still appreciated by collectors because of its beauty.

adamantine luster

Trigonal crystal system

Similar minerals
Pyrargyrite is darker and has a darker streak; cuprite has a different crystal form (mostly octahedron, cubic or rhombic dodecahedron); hutchinsonite is much softer.

prismatic crystals

Pyrargyrite, Dark Red Silver Ore
Ag_3SbS_3

Occurrences In hydro-thermal silver ore veins.

> **Hardness** 2.5–3
> **Density** 5.85
> **Luster** Metallic
> **Cleavage** Sometimes recognizable
> **Fracture** Conchoidal
> **Tenacity** Brittle

The dark red to gray-black crystals are scalalenoeder-like to prismatic. They're always translucent red. Pyrargyrite becomes darker to almost black in the light, but the actual color becomes apparent after light scratching. Like proustite (see above), pyrargyrite should always be stored in a dark place without any light.

prismatic crystals

calcite

quartz

scalenohe-dral crystal

Trigonal crystal system

Similar minerals
Proustite is brighter red; a dark tarnished proustite differs from pyrargyrite by its lighter streak; miargyrite and hutchinsonite are softer.

Crocoite, Red Lead Ore
PbCrO$_4$

Crocoite is a rare mineral. It can only occur when the elements lead and chromium, which normally do not occur together, coincide. This is the case, for example, when galena veins occur near serpentine rocks. The red crystals are acicular, tabular or prismatic and are mostly raised. A typical accompanying mineral is green pyromorphite.

Occurrences *In the oxidation zone of lead deposits.*

> **Hardness** *2.5–3*
> **Density** *5.9–6*
> **Luster** *Greasy to adamantine*
> **Cleavage** *Recognizable*
> **Fracture** *Conchoidal*
> **Tenacity** *Brittle*

Monoclinic crystal form

tabular crystals

29

limonite

prismatic crystals

striations

Similar minerals
Cinnabar has a different crystal form; realgar differs in its paragenesis; cuprite has a different crystal form.

Copper, (Native)
Cu

Occurrences *In the cementation zone of many copper deposits, often in large masses and plates (up to several tons in weight).*

> **Hardness** *2.5–3*
> **Density** *8.93*
> **Luster** *Metallic*
> **Cleavage** *None*
> **Fracture** *Hackly*
> **Tenacity** *Brittle, ductile*

Native copper often forms highly distorted, skeletal crystal aggregates and plates. The normally copper-red metallic mineral is often coated with green malachite. The typical forms, however, always show the original copper. If you scratch away the malachite in one place, you will immediately recognize the fresh copper. The coating can be removed with diluted hydrochloric acid. Native copper can be processed directly by hammering, the first evidence of its use is believed to date back to the end of the Stone Age. Today, solid copper no longer plays a role. Nowadays, the economically important copper metal (electrical cables, coin metal, etc.) is extracted from other ores.

Isometric crystal form

deformed crystals

malachite coating

Similar minerals
Silver and gold have a different color and a different streak; nickeline has a different color streak.

30

typical color of fresh copper

hackly fracture

metallic luster

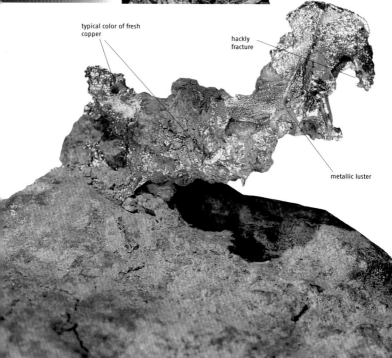

Cuprite, Red Copper Ore, Chalcotrichite
Cu₂O

Cuprite forms deep red cubic crystals but also acicular crystal druzy called chalcotrichite. The crystals are often covered with green malachite.

Occurrences *In the oxidation zone of copper deposits, especially at the boundary of the cementation zone.*

> **Hardness** *3.5*
> **Density** *6.15*
> **Luster** *Metallic, adamantine, also dull in aggregates*
> **Cleavage** *Recognizable*
> **Fracture** *Conchoidal*
> **Tenacity** *Brittle*

Isometric crystal form

Similar minerals
Hematite is harder; cinnabar has a different crystal form; cuprite is characterized by its paragenesis with malachite. Acicular cuprite differs from hutchinsonite in that it is brighter red; kermesite is much softer; solid copper is ductile; proustite has a completely different crystal form.

mat of fine acicular chalcoctrichite crystals

octahedral crystals

adamantine luster

limonite

Roselite

$Ca_2(Co,Mg)(AsO_4)_2 \cdot 2 H_2O$

Occurrences In the oxidation zone of cobalt deposits, in cavities of ores and gangue.

> **Hardness** 3.5
> **Density** 3.5–3.74
> **Luster** Vitreous
> **Cleavage** Perfect
> **Fracture** Uneven
> **Tenacity** Brittle

Monoclinic crystal form

Roselite forms dark pink to deep red, mostly lenticular, tabular crystals. Wendwilsonite, which is richer in magnesium but has the same composition, cannot be distinguished without chemical analysis.

roselite

wendwilsonite

tabular crystal

Similar minerals
The crystal form and color of roselite or wendwilsonite are very characteristic and hardly allow for confusion.

Heterosite

$(Fe,Mn)PO_4$

Fragment

Occurrences Embedded in phosphate-bearing pegmatites.

> **Hardness** 4–4.5
> **Density** 3.4
> **Luster** Vitreous, silky on cleavage planes
> **Cleavage** Good
> **Fracture** Uneven
> **Tenacity** Brittle

Orthorhombic crystal form

Heterosite almost never forms crystals, rather it occurs as embedded masses and fragments. The purple color clearly intensifies when dabbed with hydrochloric acid. Many purple heterosite pieces in collections are often mistakenly referred to as purpurite.

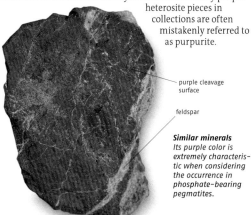

purple cleavage surface

feldspar

Similar minerals
Its purple color is extremely characteristic when considering the occurrence in phosphate-bearing pegmatites.

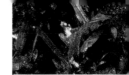

Lepidocrocite, Ruby Mica
FeOOH

Lepidocrocite forms mats of thin tabular crystals, rosettes and radial aggregates ranging in color from ruby to yellow-red. Its name refers to its color, which is said to be similar to that of the crocus flowers.

tabular crystals

Occurrences *In the oxidation zone of iron deposits, significantly rarer than goethite.*

> **Hardness** 5
> **Density** 4
> **Luster** *Adamantine*
> **Cleavage** *Perfect*
> **Fracture** *Uneven*
> **Tenacity** *Brittle*

stalactitic aggregate of flaky crystals

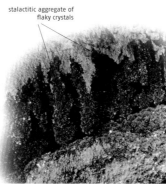

Orthorhombic crystal form

Similar minerals
Lepidocrocite differs from goethite by its red color and streak color; jarosite, natrojarosite and beudantite have a different streak color; hematite is harder.

Piemontite
Ca₂(Mn,Al)(Al₂[O/OH/SiO₄/Si₂O₇])

columnar crystals

Piemontite belongs to the epidote group. It is a so-called manganese epidote. Manganese is also responsible for its deep red to black-red color. Its crystals are mostly embedded in quartz and form fibrous to radial aggregates up to several centimeters long. It was named after where it was first discovered—near St. Marcel in Piedmont, Italy.

quartz

Occurrences *In metamorphic manganese deposits, in fissures in metamorphic rocks.*

> **Hardness** 6.5
> **Density** 3.4
> **Luster** *Vitreous*
> **Cleavage** *Poorly recognizable*
> **Fracture** *Conchoidal*
> **Tenacity** *Brittle*

parallel fibrous crystals

quartz

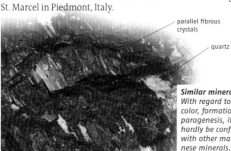

Similar minerals
With regard to color, formation and paragenesis, it can hardly be confused with other manganese minerals.

Monoclinic crystal form

Hematite, Kidney Ore, Specularite

Fe₂O₃

Fe_2O_3

Occurrences *In iron deposits.*

> **Hardness** *6.5*
> **Density** *5.2–5.3*
> **Luster** *Metallic to dull*
> **Cleavage** *None, but often flaky*
> **Fracture** *Conchoidal*
> **Tenacity** *Brittle*

Hematite forms metallic black-gray pyramidal crystals, sometimes reminiscent of octahedrons, thick to thin plates, rosette-shaped aggregates. It is also often coarse, flaky, with radials and a smooth surface (kidney ore), earthy, crusty. Beautiful crystals, especially the rosette-shaped aggregates (iron roses) can be found primarily in alpine-type fissures in the Austrian and Swiss Alps. Extraordinary hematite crystals that are rich in faces originate from the iron deposits on the Italian island of Elba.

Trigonal crystal system

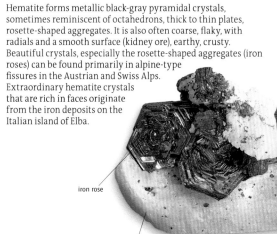

iron rose

adularia crystal

conchoidal fracture

thick tabular crystal

tarnish

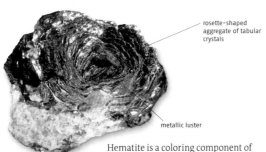

rosette-shaped aggregate of tabular crystals

metallic luster

Occurrences In iron deposits.

> **Hardness** 6.5
> **Density** 5.2–5.3
> **Luster** Metallic to dull
> **Cleavage** None, but often flaky
> **Fracture** Conchoidal
> **Tenacity** Brittle

Hematite is a coloring component of many sedimentary rocks and is also found in deposits where limestone has been converted into iron ore by ferrous solutions or gases. Clay rocks, which are characterized by high contents of finely dispersed hematite, are referred to as red chalk and are used as a color pigment (red chalk drawings). Fine-leaved silvery hematite is often called specularite.

Trigonal crystal system

Hematite with smooth, reniform aggregates is called kidney ore.

radial aggregate

Did you know?

The particularly beautiful variety of hematite, known as kidney ore, comes from Cumberland, England. This mineral is also cut into gemstones and is referred to as bloodstone. This name is due to a unique characteristic: When hematite is cut, its red streak color turns the grinding fluid red, as if the stone were bleeding.

Realgar, Red Arsenic
AsS

Monoclinic crystal form

36

Realgar forms deep red prismatic to acicular crystals and coarse embedded masses. It is extremely sensitive to light. If it is not stored in a dark place, it quickly turns orange and transforms into a yellowish powder of the same chemical composition called pararealgar. Realgar was used in the production of pesticides (e.g. rat poison), but this is no longer permitted due to its toxicity. The name "red arsenic" also indicates that the substance is harmful to health.

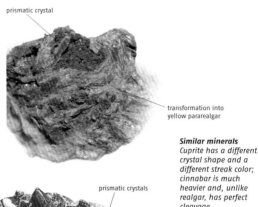

prismatic crystal

transformation into yellow pararealgar

Similar minerals
Cuprite has a different crystal shape and a different streak color; cinnabar is much heavier and, unlike realgar, has perfect cleavage.

prismatic crystals

calcite crystal

Orpiment, Yellow Arsenic
As_2S_3

radial fibers

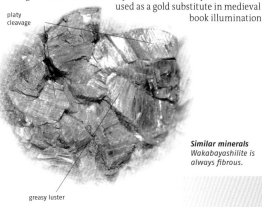

Orpiment forms prismatic, acicular, lenticular crystals and radial aggregates and crusts. Often it also forms embedded flaky aggregates. The intense yellow color of orpiment was popular as a yellow pigment for a long time, but its use was dangerous due to the toxicity of the mineral. Orpiment was used as a gold substitute in medieval book illumination.

platy cleavage

greasy luster

Occurrences *In hydro-thermal veins and in fissures and cracks in clay rocks.*

> ***Hardness*** 1.5−2
> ***Density*** 3.48
> ***Luster*** *Greasy*
> ***Cleavage*** *Completely perfect*
> ***Fracture*** *Flaky*
> ***Tenacity*** *Brittle, sectile, thin flakes of orpiment are flexible*

Monoclinic crystal form

Similar minerals
Wakabayashilite is always fibrous.

Wakabayashilite
$(As,Sb)_{11}S_{16}$

prismatic crystals

Wakabayashilite forms embedded and raised lemon-yellow acicular crystals. Because of its appearance, wakabayashilite was previously called "hair orpiment." Accompanying mineral is often red realgar. It was named after the Japanese mineralogist Yaichiro Wakaba-yashi.

calcite

Similar minerals
Orpiment is more flaky, and never as needle-like as wakabayashilite of the same color.

acicular crystals

realgar

Occurrences *In hydrothermal veins and low-temperature arsenic deposits.*

> ***Hardness*** 1.5−2
> ***Density*** 3.96
> ***Luster*** *Greasy*
> ***Cleavage*** *Completely perfect*
> ***Fracture*** *Flaky, fibrous*
> ***Tenacity*** *Brittle, sectile, thin needles of wakaba-yashilite are flexible*

Monoclinic crystal form

Beraunite

$Fe_3[(OH)_3/(PO_4)_2] \cdot 2\frac{1}{2} H_2O$

Occurrences In phosphate-bearing pegmatites resulting from weathering of primary phosphate minerals, in brown iron deposits.

> **Hardness** 3–4
> **Density** 2.9
> **Luster** Vitreous
> **Cleavage** Good, but only recognizable by larger crystals
> **Fracture** Uneven
> **Tenacity** Brittle

Monoclinic crystal form

Beraunite is formed as a secondary mineral during the transformation of primary phosphates or from the phosphate content of limonite deposits. It forms intense red tabular crystals and yellowish to green bundles of needles.

acicular (green beraunite)

limonite

radial aggregate

tabular crystals

Similar minerals Yellow beraunite is sometimes indistinguishable from struncite by simple means, but the latter is usually more clearly a straw yellow.

38

Nealite

$Pb_4Fe[Cl_2|AsO_3]_2 \cdot 2 H_2O$

Occurrences In ancient lead slags from the smelting of ores containing arsenic that were dumped into the sea.

> **Hardness** 4
> **Density** 5.88
> **Luster** Adamantine to greasy
> **Cleavage** Hardly recognizable
> **Fracture** Conchoidal
> **Tenacity** Brittle

Triclinic crystal form

So far, nealite has only been found in ancient slags in the Greek city of Laurion. There it forms tabular to prismatic yellow to yellow-brown crystals in the cavities of the slags thrown into the sea.

adamantine luster

crystal sheaf

nealite-bearing ancient slag

Similar minerals The crystal form is very typical; considering the type of occurrence, confusion is hardly possible.

Gold (Native)
Au

Gold forms octahedrons, cubes, which are rarely well formed. Much more frequently formed are dendritic, skeletal or flaky aggregates, sheets or, less frequently, wire-like formations. Gold is often coarse, embedded; you can find gold flakes, nuggets and rarely whole lumps of gold in streams and rivers. The largest gold nugget ever found weighed almost 215 kilograms and was discovered in Australia. A characteristic feature of gold is that it can be hammered into thin sheets (gold leaf) and stretched into long wires. These properties make gold the perfect metal for jewelry making, for which it has been used since the oldest advanced civilizations until today. For a long time, gold was also an important coin metal and was used for currency hedging. The biggest consumer of gold today, however, is not the jewelry industry, rather it is the electrical industry, where gold is used for contacts, plugs, etc.

Occurrences In hydro-thermal quartz veins of high to moderate temperature, in placers in rivers and streams.

> *Hardness* 2.5–3
> *Density* 15.5–19.3
> *Luster* Metallic
> *Cleavage* None
> *Fracture* Hackly
> *Tenacity* Brittle, very ductile

Isometric crystal form

39

Similar minerals
Pyrite, chalcopyrite and marcasite have a black streak and are not ductile. They are also much harder.

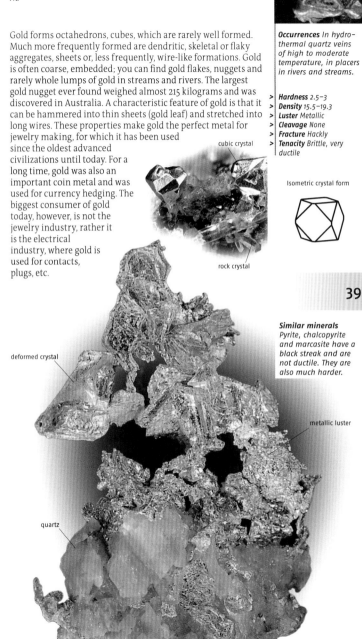

cubic crystal

rock crystal

deformed crystal

metallic luster

quartz

Cacoxenite
$Fe_4[OH/PO_4]_3 \cdot 12\ H_2O$

Occurrences In phosphate-bearing pegmatites and brown iron deposits.

> **Hardness** 3
> **Density** 2.3
> **Luster** Silky to vitreous
> **Cleavage** Not recognizable due to its thin aricular to fibrous structure
> **Fracture** Fibrous
> **Tenacity** Brittle

Cacoxenite characteristically forms golden-yellow acicular coatings, spherical and tufted aggregates, or tiny yellow globules accompanied by other phosphate minerals. In brown iron deposits, the presence of cacoxenite is a striking indication of undesirable phosphate contents in the iron ore. Hence the name given to the mineral in the 19th century. It is taken from the ancient Greek language and means "unwanted guest" or "evil stranger."

strengite

spherical aggregates

Hexagonal crystal form

Similar minerals
Struncite is pale yellow, but sometimes indistinguishable from cacoxenite by simple means.

acicular crystals

strengite

rockbridgeite

Jarosite

KFe₃(OH)₆(SO₄)₂

$KFe_3(OH)_6(SO_4)_2$

Jarosite forms brown tabular to rhombohedral crystals, crystal druzy and, often, coarse brown to yellow-brown earthy masses and crusts as well as botryoidal aggregates. The name was given to the mineral after the Barranco del Jaroso in Spain, where it was first discovered and where today the most beautiful crystals are still found.

Occurrences *In the oxidation zone of hydrothermal deposits.*

> ***Hardness*** *3–4*
> ***Density*** *3.1–3.3*
> ***Luster*** *Vitreous*
> ***Cleavage*** *Somewhat recognizable basal cleavage*
> ***Fracture*** *Uneven*
> ***Tenacity*** *Brittle*

Similar minerals
Jarosite can only be distinguished from natrojarosite by chemical analysis; beudantite is some-what harder; goethite is harder and has a different crystal form.

limonite

tabular crystals

tabular crystal

limonite

Trigonal crystal system

Natrojarosite

NaFe₃(OH)₆(SO₄)₂

$NaFe_3(OH)_6(SO_4)_2$

Natrojarosite forms brown tabular to rhombohedral crystals, crystal druzy and, often, coarse brown to yellow-brown earthy masses and crusts as well as botryoidal aggregates. Its name refers to the fact that, unlike the related jarosite, it contains sodium instead of potassium.

Occurrences *In the oxidation zone of hydrothermal deposits.*

> ***Hardness*** *3–4*
> ***Density*** *3.1–3.3*
> ***Luster*** *Vitreous*
> ***Cleavage*** *Somewhat recognizable basal cleavage*
> ***Fracture*** *Uneven*
> ***Tenacity*** *Brittle*

spherical crystal aggregates

scorodite

Similar minerals
Natrojarosite can only be distinguished from jarosite by chemical analysis; beudantite is somewhat harder; goethite is harder and has a different crystal form.

limonite

tabular crystals

Trigonal crystal system

Beudantite

$PbFe_3[(OH)_6/SO_4/AsO_4]$

Occurrences In the oxidation zone of lead-bearing deposits, which also contain arsenic-bearing primary minerals.

> **Hardness** 4
> **Density** 4.3
> **Luster** Vitreous
> **Cleavage** None
> **Fracture** Conchoidal
> **Tenacity** Brittle

Beudantite forms rhombohedral, including pointed rhombohedral, but also cube-like crystals of yellowish to brownish, rarely olive green color. In addition, it also forms tabular crystal druzy and crusty, earthy aggregates. The best beudantite crystals were found in the Tsumeb deposit in Namibia.

thick tabular crystals

flaky crystals

Trigonal crystal system

Similar minerals
Jarosite and natro-jarosite are softer; tsumcorite is harder and has a different crystal form.

Zincite

ZnO

Occurrences In metamorphic zinc-manganese deposits, in the oxidation zone of zinc deposits, in volcanic exhalations.

> **Hardness** 4
> **Density** 5.66
> **Luster** Vitreous
> **Cleavage** Perfect
> **Fracture** Uneven
> **Tenacity** Brittle

Zincite sometimes forms pyramidal crystals, of which both crystals are formed differently (hemimorphic); much more frequently it forms fragments, granular aggregates and coarse adhesions. Zincite's defining characteristic is its mostly deep-red color. Zincite has a very high zinc content, to which it also owes its name. But because of its rarity it can virtually never be used as zinc ore.

willemite

cleavage plane

red fragment

Hexagonal crystal form

Similar minerals
Sphalerite and wurtzite are usually darker brown and do not have a yellow streak.

franklinite

willemite

Pucherite

$Bi_2V_2O_8$

Pucherite forms reddish brown to yellowish, thick tabular to isometric, and, rarely, acicular crystals that are always raised. It also forms inconspicuous crusty, earthy coatings. It is typically associated with other bismuth minerals such as bismite, bismuth or eulytine.

Occurrences In the oxidation zone of deposits of the bismuth–cobalt–nickel formation.

> **Hardness** 4
> **Density** 6.25
> **Cleavage** Perfect
> **Luster** Greasy
> **Fracture** Conchoidal
> **Tenacity** Brittle

limonite

crystal druzy

thick tabular crystals

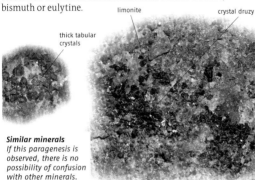

Orthorhombic crystal form

Similar minerals
If this paragenesis is observed, there is no possibility of confusion with other minerals.

Tsumcorite

$PbZnFe(AsO_4)_2 \cdot H_2O$

quartz crystal

Tsumcorite forms short prismatic to tabular crystals, flaky radial aggregates, earthy crusts. It was named after the Tsumeb Corporation, the mining company at the first discovery site in Tsumeb, Namibia, from which the only good crystal specimens originate. At other sites, tsumcorite forms only very fine crusts or powdery masses and can hardly be identified.

tabular crystals

Occurrences In the oxidation zone of lead and zinc-bearing deposits.

> **Hardness** 4.5
> **Density** 5.2
> **Luster** Vitreous
> **Cleavage** None recognizable
> **Fracture** Uneven
> **Tenacity** Brittle

crystal druzy

weathered tetrahedrite

Similar minerals
Mimetite has a different crystal form; beudantite is usually browner, but sometimes cannot be distinguished by simple means.

Monoclinic crystal form

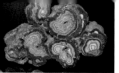

Sphalerite, Zinc Blende
ZnS

Occurrences *In ore deposits.*

> **Hardness** 3.5–4
> **Density** 3.9–4.2
> **Luster** Semi-metallic adamantine
> **Cleavage** Perfect rhombic dodecahedral
> **Fracture** Conchoidal, splintery
> **Tenacity** Brittle

Isometric crystal form

Sphalerite as zinc ore is often splintery with shiny cleavage surfaces and coarse. In addition, it often forms raised crystals, mainly tetrahedra, rhombic dodecahedron, often, octahedron-like due to the combination of two tetrahedra. Its surfaces are striated and frequently show formation of twins. The color varies from colorless to yellow through to red, brown and black. The name "zinc blende" comes from the metal content and the special, almost diamond-like shine of the mineral. Today, sphalerite is the most important zinc ore. But for some time, it could not be smelted because the zinc oxide produced during roasting quickly becomes gaseous. Therefore, brass production was not carried out using sphalerite but through its weathered products such as zinc spar, hemimorphite and hydrozincite. Their mixture was called calamine.

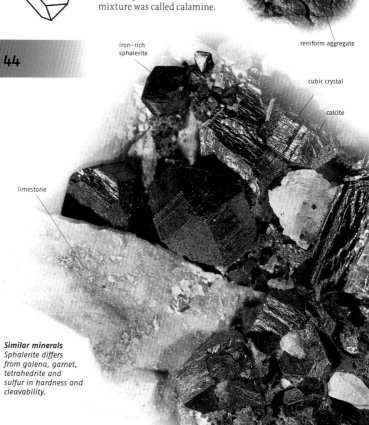

barite

reniform aggregate

iron–rich sphalerite

cubic crystal

calcite

limestone

Similar minerals
Sphalerite differs from galena, garnet, tetrahedrite and sulfur in hardness and cleavability.

Wurtzite

ZnS

Brown wurtzite sometimes forms crystals which are usually tabular, spindle-shaped or pyramidal. Much more frequently it is radial and fibrous. Though wurtzite is much rarer than sphalerite, despite having the same chemical composition, and can only be distinguished from sphalerite by its other crystal form. Because of its rarity, wurtzite is of zero importance in the mining of zinc ore.

tabular crystals

pyramidal crystal

calcite

Occurrences *In hydro-thermal veins and zinc ore deposits.*

> **Hardness** 3.5–4
> **Density** 4
> **Luster** Resinous
> **Cleavage** Basal and prismatic
> **Fracture** Uneven
> **Tenacity** Brittle

Similar minerals
Sphalerite has a different crystal form and cleavability, but can also be reniform and radial; the schalenblende variety is always concentric.

Hexagonal crystal form

Descloizite

Pb(Zn,Cu)[OH/VO$_4$]

Descloizite forms prismatic, rarely tabular crystals, often with dendritic, radial aggregates. Often crusty, coarse, radial aggregates. The color is usually a shiny and resinous brown, much rarer are red or yellow crystals. Descloizite forms through weathering even when the vanadium content of the ore is low. More abundant occurrences were mined as vanadium ore.

Occurrences *In the oxidation zone of lead deposits. The vanadium mostly comes from black shale in the surrounding area.*

> **Hardness** 3.5
> **Density** 5.5–6.2
> **Luster** Resinous
> **Cleavage** None
> **Fracture** Uneven
> **Tenacity** Brittle

limonite

acicular crystals

tabular crystals

Similar minerals
Magnetite is harder; brown calcite or smithsonite are lighter and show a clear cleavage; wulfenite has a different streak color.

Monoclinic crystal form

Keckite

$Ca(Mn,Zn)_2Fe_3(OH)_3(PO_4)_4 \cdot 2H_2O$

Occurrences *In pegmatites as a conversion product of rockbridgeite and other phosphates.*

> **Hardness** 4
> **Density** 2.7–2.9
> **Luster** Vitreous
> **Cleavage** Good
> **Fracture** Uneven
> **Tenacity** Brittle

Keckite forms yellowish to brownish coatings on rockbridgeite, often intercalated aggregates or radial, fibrous aggregates; transformed rockbridgeite fibers and coarse granular masses. Keckite was named after the Bavarian mineral collector Erich Keck.

rockbridgeite

Monoclinic crystal form

rockbridgeite   granular aggregates

Similar minerals
The minerals of the jahnsite group are indistinguishable from keckite without chemical analysis, but their occurrence during the transformation of rockbridgeite is very characteristic.

Frondelite

$(Mn,Fe)Fe_4[(OH)_5/(PO_4)_3]$

Occurrences *In phosphate-bearing pegmatites as a conversion product of primary phosphate minerals.*

> **Hardness** 4.5
> **Density** 3.4
> **Luster** Vitreous
> **Cleavage** Present, but rarely recognizable
> **Fracture** Uneven
> **Tenacity** Brittle

Frondelite rarely forms brown, prismatic to tabular crystals. More frequently, it forms radial, fibrous aggregates with a botryoidal-like structure, as well as reniform, crusty, coarse habits.

Orthorhombic crystal form

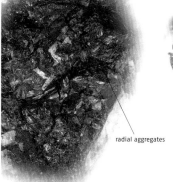

acicular crystals

Similar minerals
Color and streak color are very characteristic. If the typical occurrence is taken into account, confusion is seldom possible.

radial aggregates

Hausmannite

Mn₃O₄

Mn_3O_4

Hausmannite often forms octahedron-like iron-black crystals, sometimes with quintuplet expitaxial overgrowth. More frequently it is granular and coarse. In manganese ore deposits, it is mined together with other minerals such as brownite or manganite as manganese ore.

Occurrences *In metamorphic manganese deposits, in hydrothermal manganese ore veins.*

> **Hardness** 5.5
> **Density** 4.7–4.8
> **Luster** Metallic
> **Cleavage** Perfect basal
> **Fracture** Uneven
> **Tenacity** Brittle

quintuplet

bipyramidal crystals

Tetragonal crystal form

uneven fracture

Similar minerals
Magnetite has a black streak; manganite and pyrolusite have a different crystal form.

Manganite

MnOOH

$MnOOH$

Manganite forms long to short prismatic, rarely tabular brown-black to black crystals, rarely cruciform twins. More often, it is radial, earthy and coarse. The best crystal specimens in the world, with large and highly lustrous crystals, have been found in the Ilfeld deposit in the Harz Mountains.

Occurrences *In hydrothermal veins together with other manganese ores.*

> **Hardness** 4
> **Density** 4.3–4.4
> **Luster** Metallic
> **Cleavage** Discernible
> **Fracture** Uneven
> **Tenacity** Brittle

prismatic crystals

barite

prismatic crystal

Monoclinic crystal form

Similar minerals
Goethite has a different color; pyrolusite, in contrast to manganite, has a pure black streak and a greater hardness.

Goethite, Brown Iron Ore, Limonite
FeOOH

Occurrences Crystals in cavities formed by bubbles in volcanic rocks in the oxidation zones of various ore deposits.

> **Hardness** 5–5.5
> **Density** 4.3
> **Luster** Metallic to dull
> **Cleavage** Perfect, but only recognizable in good crystals
> **Fracture** Uneven
> **Tenacity** Brittle

The yellow, brown to blackish goethite forms acicular, prismatic and long tabular crystals; its aggregates are radial, reniform with a smooth surface (brown botryoidal limonite), coarse, earthy (limonite). Limonite often forms in the main part of the oxidation zone. Its cavities contain many, mostly colorful, oxidation minerals. Goethite is extremely widespread worldwide and often serves as a locally important iron ore. Its name honors the German poet Johann Wolfgang von Goethe, who also had a keen interest in minerals.

long tabular crystals

quartz

Orthorhombic crystal form

stalagmitic aggregate

48

Similar minerals
Lepidocrocite is clearly redder and mostly flaky; kidney ore has a red streak; psilomelane has a black streak.

reniform surface

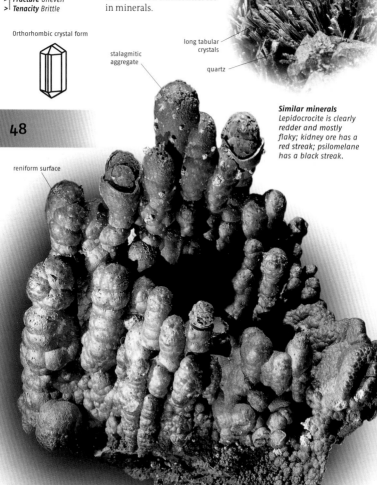

Ferberite

(Fe,Mn)WO₄

The iron-rich ferberite forms black tabular to prismatic crystals, it can also form acicular, radial, splintery, coarse habits. Ferberite is a member of a solid solution series with the two endmembers being ferberite (Fe WO4) and hubnerite (Mn WO4). It is an important tungsten ore, the best crystals come from Portugal and Kazakhstan. The mineral was named after the collector Moritz Rudolph Ferber from Gera, Germany.

parallel tabular crystal aggregate

quartz

tabular crystal

Occurrences In granites, pegmatites, pneumatolytic and hydrothermal veins.

> *Hardness* 5–5.5
> *Density* 7.14–7.54
> *Luster* Greasy metallic
> *Cleavage* Very good
> *Fracture* Uneven
> *Tenacity* Brittle

Similar minerals
Columbite is somewhat harder and does not have such good cleavability; cassiterite has a different crystal form; hubnerite is recognizably reddish.

fluorite

metallic luster

Monoclinic crystal form

Hubnerite

(Mn,Fe)WO₄

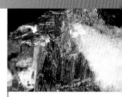

The manganese-rich hubnerite forms brown, reddish translucent tabular to prismatic crystals, but also acicular, radial aggregates and coarse masses. Hubnerite is the manganese-rich member of a solid solution series with the two endmembers being ferberite (FeWO4) and hubnerite (MnWO4). It is much rarer than ferberite and, like ferberite, is mined as tungsten ore.

Occurrences In granites, pegmatites, pneumato-lytic and hydrothermal veins.

> *Hardness* 5–5.5
> *Density* 7.14–7.54
> *Luster* Greasy metallic
> *Cleavage* Very good
> *Fracture* Uneven
> *Tenacity* Brittle

radial appearance

thick tabular crystal

Similar minerals
Ferberite is always black, never reddish; columbite and manganotantalite are found in other parageneses.

feldspar

Monoclinic crystal form

Chromite
(Fe,Mg)Cr$_2$O$_4$

Occurrences Embedded as grains and crystals in basic rocks such as peridotite, anorthosite and serpentinite.

> *Hardness* 5.5
> *Density* 4.5–4.8
> *Luster* Metallic to greasy
> *Cleavage* None
> *Fracture* Conchoidal
> *Tenacity* Brittle

Black chromite seldom forms octahedral crystals, it is usually granular, coarse, often in the form of roundish grains embedded in the rock (leopard ore). Due to its hardness and chemical resistance it is often found in placer deposits in the form of rounded grains. Chromite is the most important chromium ore. Chromium is used as an alloy metal in steel production.

Isometric crystal form

chromite crystals

crystal grain

serpentine

serpentine

Similar minerals Magnetite has a black streak and is clearly magnetic, augite has good cleavability.

Nickeline, Niccolite
NiAs

Occurrences In hydro-thermal ore veins, in gabbro.

> *Hardness* 5.5
> *Density* 7.8
> *Luster* Metallic
> *Cleavage* Mostly indiscernible
> *Fracture* Uneven
> *Tenacity* Brittle

The metallic pink niccolite rarely forms crystals, pyramids and spindle-like crystals. It is almost always coarse, with radials or forms reniform aggregates. Together with other nickel sulfides and nickel arsenides, nickeline serves as an important nickel ore. Nickel is especially used as an alloy metal in the production of steel.

thick tabular crystals

metallic luster

Hexagonal crystal form

Similar minerals The much rarer maucherite is somewhat lighter, otherwise nickeline is unmistakable because of its color; pyrite is yellower and harder; pyrrhotite has a black streak.

crystal aggregate

calcite

Neptunite

Na₂FeTi[Si₄O₁₂]

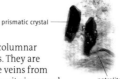

prismatic crystal

Black neptunite forms prismatic, columnar
crystals that are often rich in faces. They are
mostly embedded, e.g., in natrolite veins from
which they can be extracted. Neptunite is named
after Neptune, the Roman god of the sea because of its
occurrence together with aegirine,
named after the Nordic sea god,
Aegir.

natrolite

prismatic crystals

natrolite

*Occurrences Crystals
embedded in alkaline
pegmatites, natrolite
veins.*

> **Hardness** 5.5
> **Density** 3.23
> **Luster** *Vitreous*
> **Cleavage** *Mostly unrecog-
> nizable*
> **Fracture** *Conchoidal*
> **Tenacity** *Brittle*

Similar minerals
*Tourmaline has a
distinctly different
crystal form and is
harder; aegirine has
a typical cleavage
angle of 90°; the
same applies to
augite; hornblende
has a cleavage angle
of 120°.*

Monoclinic crystal form

Aeschynite-(Ce)

(Ce,Th,Ca)(Ti,Nb,Ta)₂O₆

Aeschynite that is embedded as tabular crystals always has a
black pitchy luster. The acicular to long tabular raised crystals
are brownish in color. The name aeschynite comes from the
Greek word for "shame." Early examiners of the mineral said
it would be a shame to reveal its interior because it was so
difficult to analyze.

quartz

pitchy luster,
conchoidal
fracture

thick tabular
crystal

*Occurrences Embedded
in granite pegmatites,
raised in alpine-type
fissures.*

> **Hardness** 5–6
> **Density** 4.9–5.1
> **Luster** *Pitchy (embedded),
> vitreous (raised)*
> **Cleavage** *None*
> **Fracture** *Conchoidal*
> **Tenacity** *Brittle*

Orthorhombic crystal form

Similar minerals
*Rutile has a tetragonal
symmetry; raised
allanite crystals are
more purple and, like
samarskite, have a
different streak color.*

Hornblende
$(Ca,Na,K)_{2-3}(Mg,Fe,Al)_5[(OH,F)_2/(Si,Al)_2Si_6O_{22}]$

Occurrences
*Rock-forming
component in
magmatic-volcanic and
metamorphic rocks.*

> **Hardness** *5–6*
> **Density** *2.9–3.4*
> **Luster** *Vitreous to greasy*
> **Cleavage** *Perfect, the
> cleavage surfaces form an
> angle of about 120°*
> **Fracture** *Uneven*
> **Tenacity** *Brittle*

Basaltic hornblende is always black and
forms embedded crystals in volcanic
rocks; particularly beautiful, fully formed
crystals in volcanic tuffs. Common hornblende is black to green, usually radially
embedded, less often in splintery masses.

basaltic hornblende

prismatic crystal

Monoclinic crystal form

radial aggregate

acicular crystals

mica

Similar minerals
*Augite has a different
cleavage angle than
hornblende; tourma-
line has no cleavage;
neptunite has a
different crystal form
and no cleavage.*

Franklinite

$ZnFe_2O_4$

Franklinite usually forms black embedded octahedrons together with other zinc-rich minerals, such as willemite or zincite. It rarely forms raised crystals and is often granular and coarse. The best and largest crystals come from the Franklin Mine in New Jersey, after which the mineral was named.

Occurrences In meta-morphic zinc deposits.

> *Hardness* 6–6.5
> *Density* 5.0–5.2
> *Luster* Metallic
> *Cleavage* None
> *Fracture* Conchoidal
> *Tenacity* Brittle

octahedron

willemite

Isometric crystal form

octahedral crystal

calcite

willemite

calcite

Similar minerals
Franklinite differs from magnetite by its paragenesis with zinc minerals; thin splinters of gahnite are always somewhat greenish.

Babingtonite

$Ca_2FeFeSi_5O_{14}OH$

Occurrences In fissures in granite, pegmatites and cavities in volcanic rocks.

> ***Hardness*** 5.5–6
> ***Density*** 3.25–3.35
> ***Luster*** Vitreous
> ***Cleavage*** Perfect
> ***Fracture*** Uneven
> ***Tenacity*** Brittle

Triclinic crystal form

Babingtonite forms black thick tabular to short prismatic crystals that are mostly raised. It is seldom crudely ingrown.

thick tabular crystals

tabular crystal

quartz

Similar minerals
Axinite has a different streak and is usually brighter and has much sharper-edged crystals; augite and diopside, in contrast to babingtonite, have a cleavage angle of 90°; hornblende has a cleavage angle of 120°.

Helvine

$(Fe,Mn,Zn)_8[S_2/(BeSiO_4)_6]$

Occurrences In skarn deposits.

> ***Hardness*** 6
> ***Density*** 3.1–3.66
> ***Luster*** Vitreous
> ***Cleavage*** None
> ***Fracture*** Conchoidal
> ***Tenacity*** Brittle

Isometric crystal form

Helvine forms yellow to brown, raised and embedded, tetrahedra. Rarely is it rhombic dodecahedron, and often it is coarse.

tetrahedral crystals

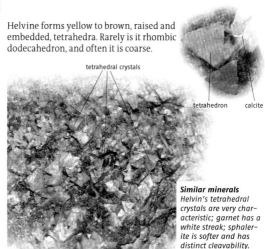

tetrahedron

calcite

Similar minerals
Helvin's tetrahedral crystals are very characteristic; garnet has a white streak; sphalerite is softer and has distinct cleavability.

Rutile
TiO₂

Rutile crystals form prismatic to capillary habits and are yellow, red or black in color. Twins are often of different natures; knee-shaped twins with a flat angle and heart-shaped twins with an acute angle between the two crystals. If both twins appear at the same time, the so-called sagenite grids form, which are completely typical for rutile. Artificially produced rutile (titanium dioxide) has great economic significance as a white pigment in wall paints, as UV protection in sunscreens or as an impregnating agent in the manufacture of UV-impermeable children's clothing.

Occurrences In pegmatites, in alpine–type fissures, in sedimentary rocks, metamorphic rocks and placers.

> *Hardness* 6
> *Density* 4.2–4.3
> *Luster* Adamantine to metallic
> *Cleavage* Perfect, but only visible in thick crystals
> *Fracture* Conchoidal
> *Tenacity* Brittle

Tetragonal crystal form

55

sagenite

chlorite

hematite

acicular crystals

Similar minerals
Tourmaline is harder and has a different luster; magnetite has a different streak; brookite and anatase have a different crystal form.

Tyrolite

$Ca_2Cu_9[(OH)_{10}/(AsO_4)_4] \cdot 10\ H_2O$

Occurrences In the oxidation zone of copper deposits.

> **Hardness** 2
> **Density** 3.2
> **Luster** Pearly
> **Cleavage** Very perfect basal
> **Fracture** Flaky
> **Tenacity** Brittle, flexible sheets

Orthorhombic crystal form

Tyrolite forms greenish blue thin tabular crystals and flakes, often forming rosettes, spherical aggregates and coarse and crusty coatings.

radial aggregate

azurite

platy cleavage

pearly luster

Similar minerals
Brochantite is pure green; is darker blue; chalcophyllite has a different crystal form; clinotyrolite is indistinguishable by simple means.

Clinotyrolite

$Ca_2Cu_9[(OH)_{10}/(AsO_4)_4] \cdot 10\ H_2O$

Occurrences In the oxidation zone of copper deposits.

> **Hardness** 2
> **Density** 3.2
> **Luster** Pearly
> **Cleavage** Very perfect basal
> **Fracture** Flaky
> **Tenacity** Brittle, flexible sheets

Monoclinic crystal form

Clinotyrolite forms greenish blue thin tabular crystals and flakes, often forming rosettes or spherical aggregates and coarse and crusty coatings.

radial flaky aggregate

tabular crystals

pearly luster

calcite

Similar minerals
Brochantite is pure green; azurite is darker blue; chalcophyllite has a different crystal form; tyrolite is indistinguishable by simple means.

Ktenasite

$(Cu,Zn)_3(SO_4)(OH)_4 \cdot 2\,H_2O$

Ktenasite rarely forms greenish blue to green tabular crystals and flakes. More often it forms rosettes or spherical aggregates and coarse and crusty coatings.

Occurrences In the oxidation zone of copper-zinc deposits.

> **Hardness** 2–2.5
> **Density** 2.9
> **Luster** Vitreous
> **Cleavage** Recognizable
> **Fracture** Uneven
> **Tenacity** Brittle

tabular crystal

gypsum

flaky crystal aggregate

spherical crusts

Similar minerals
Brochantite is pure green; azurite is darker blue; chalcophyllite has a different crystal form; clinotyrolite and tyrolite tend to have thinner flakes and are more blue.

Monoclinic crystal form

Chalcophyllite

$Cu_{18}Al_2[(OH)_{27}/(AsO_4)_3/(SO_4)_3] \cdot 36\,H_2O$

Chalcophyllite forms blue-green to emerald-green thin tabular six-sided flakes, rosettes, crusts and coatings on limonite and corroded surrounding rock.

Occurrences In the oxidation zone of copper deposits.

> **Hardness** 2
> **Density** 2.67
> **Luster** Vitreous
> **Cleavage** Perfect
> **Fracture** Flaky
> **Tenacity** Flexible

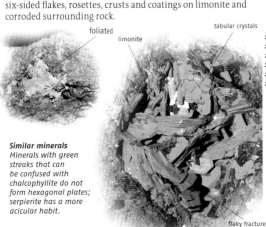

foliated

limonite

tabular crystals

flaky fracture

Similar minerals
Minerals with green streaks that can be confused with chalcophyllite do not form hexagonal plates; serpierite has a more acicular habit.

Hexagonal crystal form

Chlorite
$(Fe,Mg,Al)_6[(OH)_2/(Si,Al)_4O_{10}]$

Occurrences *In meta-morphic rocks (chlorite schist) and sediments.*

> **Hardness** 2
> **Density** 2.6–3.3 (depend-ing on iron content)
> **Luster** Vitreous, pearly on cleavage planes
> **Cleavage** Perfect basal
> **Fracture** Flaky
> **Tenacity** Brittle, flexible

Chlorite forms green to black-green tabular crystals, sheets, spherical and vermicular aggregates, crusts and powdery to sandy masses. It also forms whole rocks as the main mixture part (see chlorite schists). In meta-morphic rocks, such as gneiss, it is a typical indicator of the occurrence of alpine-type fissures.

vermicular aggregates

pearly luster

platy cleavage

quartz

Monoclinic crystal form

Similar minerals
Mica is harder and elastic.

58

platy cleavage

tablet crystal aggregate of low iron chlorite

Chrysocolla

CuSiO₃+aq.

Chrysocolla does not form crystals. It can be found as botryoidal, reniform masses, radial aggregates, crusty, stalactitic and coarse — often in the form of other copper minerals, e.g. azurite. Because of its beautiful green to blue color and the varied patterning of the aggregates, it is used as a gemstone (cabochon). Its name translates as "gold glue," since it was formerly used by goldsmiths as an aid in the production of small gold spheres, for example.

cabochon

Occurrences In the oxidation zone of copper deposits.

> *Hardness 2–4*
> *Density 2–2.2*
> *Luster Vitreous, slightly greasy*
> *Cleavage None*
> *Fracture Conchoidal*
> *Tenacity Brittle*
> *Crystal form Mostly amorphous*

Similar minerals Malachite has a different color; turquoise is harder.

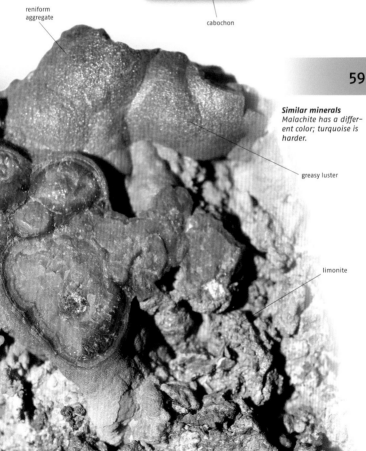

reniform aggregate

greasy luster

limonite

Olivenite
$Cu_2[OH/AsO_4]$

prismatic crystals

Occurrences In the oxidation zone of copper
deposits.

> **Hardness** 3
> **Density** 4.3
> **Luster** Vitreous to silky
> **Cleavage** None
> **Fracture** Conchoidal
> **Tenacity** Brittle

Olivenite forms dark to light green
tabular, prismatic to acicular crystals
and radial aggregates. Very fine-grained
aggregates can be practically white.

radial aggrega

Orthorhombic crystal form

Similar minerals
Adamite is usually
much brighter
green, but in the
copper-containing
variety cuproadamite
is often difficult to distinguish; the same ap-
plies to libethenite, but
paragenesis with other
arsenic-containing
minerals indicates the
presence of olivenite.

quartz

Libethenite
$Cu_2[OH/PO_4]$

bipyramidal
crystals

Occurrences In the oxidation zone of copper
deposits.

> **Hardness** 3
> **Density** 4.3
> **Luster** Vitreous
> **Cleavage** None
> **Fracture** Conchoidal
> **Tenacity** Brittle

Libethenite forms green to black-green
crystals that are acicular and prismatic as
well as octahedron-like. It also forms
radials, botryoidal or reniform aggregates
and crusty and coarse coatings.

isometric crystals

limonitic quartz

Monoclinic crystal form

Similar minerals
Adamite is usually much
brighter green, but in
the copper-containing
variety cuproadamite
is often difficult to
distinguish; the same
applies to olivenite, but
paragenesis with other
phosphorous-containing
minerals indicates the
presence of libethenite.

calcite

Atacamite

$Cu_2(OH)_3Cl$

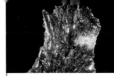

tabular crystal

radial appearance

Atacamite forms prismatic to acicular crystals, more rarely tabular or octahedron-like crystals. It is often found as radial rosettes, flaky,

prismatic crystals

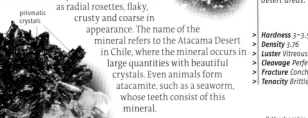

crusty and coarse in appearance. The name of the mineral refers to the Atacama Desert in Chile, where the mineral occurs in large quantities with beautiful crystals. Even animals form atacamite, such as a seaworm, whose teeth consist of this mineral.

Occurrences *In the oxidation zone of copper deposits, especially in desert areas.*

> ***Hardness*** 3–3.5
> ***Density*** 3.76
> ***Luster*** Vitreous
> ***Cleavage*** Perfect
> ***Fracture*** Conchoidal
> ***Tenacity*** Brittle

Orthorhombic crystal form

olivenite

Similar minerals
Malachite effervesces when dabbed with hydrochloric acid; brochantite is somewhat harder and not so blackish green.

Mottramite

$Pb(Cu,Zn)[OH/VO_4]$

Mottramite rarely forms prismatic crystals, usually found in olive-green to black-green radial, crusty and reniform aggregates, dendritic, as crusty earthy coatings. Mottramite is rarer than the related zinc-rich

reniform aggregates

descloicite and is used together with it as a vanadium ore.

radial appearance

Occurrences *In the oxidation zone of vanadium-rich lead and copper deposits.*

> ***Hardness*** 3.5
> ***Density*** 5.7–6.2
> ***Luster*** Resinous
> ***Cleavage*** None
> ***Fracture*** Uneven
> ***Tenacity*** Brittle

reniform

calcite

Similar minerals
Descloizite is more brown; malachite effervesces when dabbed with hydrochloric acid and is more emerald green; ktenasite is bluer; brochantite is clearly emerald green, as is atacamite.

Orthorhombic crystal form

Agardite

$(Ce,Y,La,Ca)_2Cu_{12}[(OH)_{12}/(AsO_4)_6] \cdot 6\ H_2O$

Occurrences *In the oxidation zone of copper deposits.*

> **Hardness** 3–4
> **Density** 3.6–3.7
> **Color** Yellowish green to bluish green
> **Streak color** Pale green
> **Luster** Vitreous
> **Cleavage** None recognizable
> **Fracture** Uneven

Hexagonal crystal form

Agardites are a group of minerals that all form yellowish to blue-green acicular, capillary crystal bundles. Depending on the predominant rare-earth element, they are called agardite-(Ce), agardite-(Y) or agardite-(La).

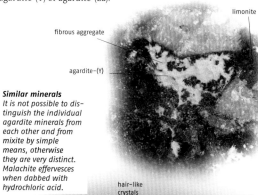

limonite

fibrous aggregate

agardite–(Y)

Similar minerals
It is not possible to distinguish the individual agardite minerals from each other and from mixite by simple means, otherwise they are very distinct. Malachite effervesces when dabbed with hydrochloric acid.

agardite–(La)

hair-like crystals

limonite

Brochantite
$Cu_4[(OH)_6/SO_4]$

Brochantite forms emerald green acicular, rarely tabular, crystals, radials, fibrous, reniform aggregates, velvety coatings. Often it is granular, earthy.

Occurrences In the oxidation zone of copper deposits.

acicular crystals

prismatic crystals

> **Hardness** 3.5–4
> **Density** 3.97
> **Luster** Vitreous, pearly on cleavage planes
> **Cleavage** Perfect, but mostly unrecognizable because of the acicular crystals
> **Fracture** Uneven
> **Tenacity** Brittle

Monoclinic crystal form

Similar minerals
Malachite effervesces when dabbed with hydrochloric acid; atacamite is softer and usually somewhat darker.

Euchroite
$Cu_2AsO_4OH \cdot 3\ H_2O$

blocky crystals

Euchroite forms short prismatic to thick tabular crystals of emerald green to grass-green color, which are almost always raised.

thick tabular crystals

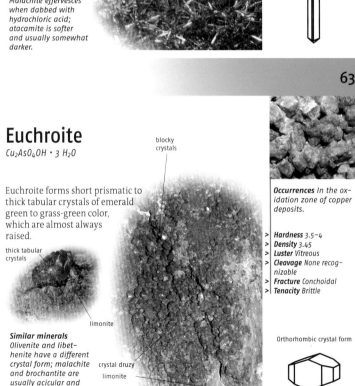

limonite

crystal druzy
limonite

Occurrences In the oxidation zone of copper deposits.

> **Hardness** 3.5–4
> **Density** 3.45
> **Luster** Vitreous
> **Cleavage** None recognizable
> **Fracture** Conchoidal
> **Tenacity** Brittle

Orthorhombic crystal form

Similar minerals
Olivenite and libethenite have a different crystal form; malachite and brochantite are usually acicular and more emerald green; olivenite is more olive green.

Malachite
$Cu_2[(OH)_2/CO_3]$

Occurrences In the oxidation zone of copper deposits.

> **Hardness** 4
> **Density** 4
> **Luster** Vitreous, silky in aggregates, also dull
> **Cleavage** Good, but practically indiscernible due to the mostly acicular or fibrous habit
> **Fracture** Conchoidal
> **Tenacity** Brittle

Monoclinic crystal form

Similar minerals
Confusable minerals do not effervesce when dabbed with diluted hydrochloric acid.

Malachite is the most common oxidation mineral in the oxidation zone of copper deposits. There, it forms emerald green acicular to tabular crystals, tufted to radial crusts and reniform aggregates. It is often found as a crust on cuprite and solid copper. Azurite crystals are often transformed into malachite, while the typical azurite form is preserved.

acicular crystals

limonite

prismatic crystals

radial aggregate

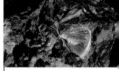

Malachite is also used as a gemstone because of its beautiful green color and typical light and dark-green bands. Cabochons for brooches and pendants and beads for stone necklaces are cut from it. Malachite is often used to make handcrafted items.

Occurrences In the oxidation zone of copper deposits.

> **Hardness** 4
> **Density** 4
> **Luster** Vitreous, silky in aggregates, also dull
> **Cleavage** Good, but practically indiscernible due to the mostly acicular or fibrous habit
> **Fracture** Conchoidal
> **Tenacity** Brittle

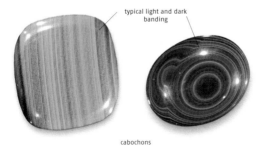

typical light and dark banding

cabochons

Monoclinic crystal form

Distinction
The almost always present bands make the stone very typical and unmistakable in appearance. Chrysocolla is always clearly a bluish color.

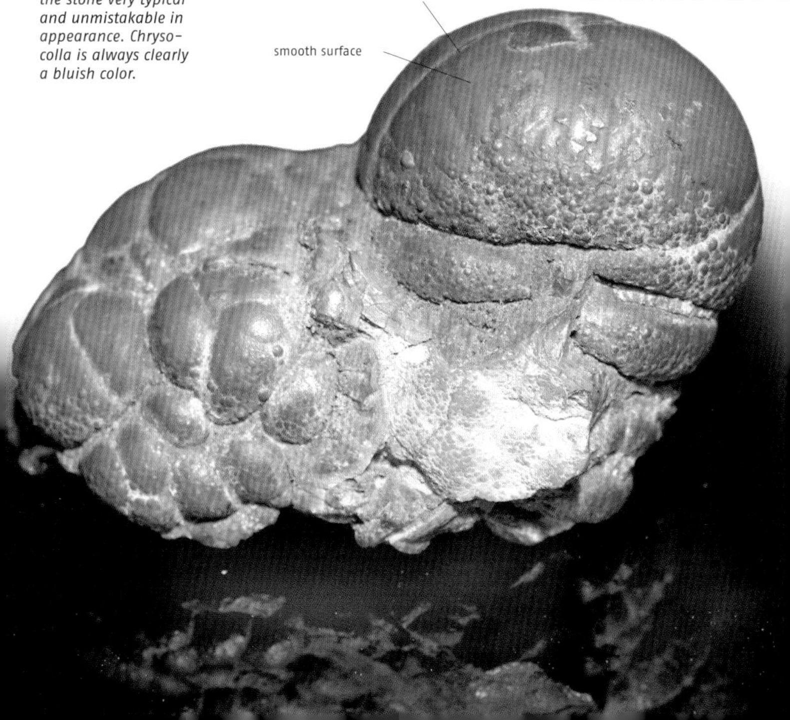

reniform aggregate

smooth surface

Pseudomalachite
Cu₅[(OH)₂/PO₄]₂

$Cu_5[(OH)_2/PO_4]_2$

Occurrences *In the oxidation zone of copper deposits.*

> **Hardness** 4.5
> **Density** 4.34
> **Luster** *Vitreous to greasy*
> **Cleavage** *None*
> **Fracture** *Conchoidal*
> **Tenacity** *Brittle*

Monoclinic crystal form

Tabular crystals of pseudomalachite are rather rare, mostly forming emerald to blackish green radial and reniform aggregates, crusty, earthy coatings.

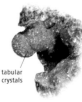
tabular crystals

smooth surface
reniform aggregate

Similar minerals
In contrast to pseudomalachite, malachite effervesces when dabbed with diluted hydrochloric acid; cornwallite cannot be distinguished by simple means, but the paragenesis of pseudomalachite together with phosphorus–containing minerals is always indicative.

Conichalcite
CaCu[OH/AsO₄]

$CaCu[OH/AsO_4]$

Occurrences *In the oxidation zone of copper deposits.*

> **Hardness** 4.5
> **Density** 4.33
> **Luster** *Vitreous*
> **Cleavage** *None recognizable*
> **Fracture** *Uneven*
> **Tenacity** *Brittle*

Orthorhombic crystal form

Conichalcite rarely forms prismatic crystals, rather it typically forms apple-green radial, reniform aggregates and warty, crusty, coarse coatings. Conichalcite is always formed when the surrounding rock is rich in calcium (e.g., limestone, marble).

limonite

crystal crust
spherical crystal aggregate

Similar minerals
The apple green color is very characteristic and distinguishes conichalcite from malachite, oliveite, cuproadamite or cornwallite. Scorodite can be greenish, but shows a different crystal form.

Bayldonite
$PbCu_3[OH/AsO_4]_2$

Bayldonite forms green to yellow-green thick tabular crystals, pseudohexagonal triplets, as well as crusty and radial aggregates. Formed by the transformation of mimetesite, inside often still has the form of mimetesite (pseudomorphosis).

Occurrences In the oxidation zone of lead-copper deposits.

> **Hardness** 4.5
> **Density** 5.5
> **Luster** Resinous
> **Cleavage** None
> **Fracture** Uneven
> **Tenacity** Brittle

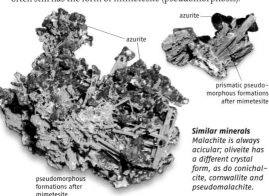

azurite

azurite

prismatic pseudo-morphous formations after mimetesite

pseudomorphous formations after mimetesite

Monoclinic crystal form

Similar minerals
Malachite is always acicular; oliveite has a different crystal form, as do conichalcite, cornwallite and pseudomalachite.

Chalcosiderite
$CuFe_6[(OH)_8/(PO_4)_4] \cdot 4\ H_2O$

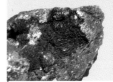

Chalcosiderite sometimes forms dark green short prismatic to thick tabular crystals, lighter green crusts and coatings and also has coarse occurrences. It forms whenever there is sufficient phosphorus in the deposit.

Occurrences In the oxidation zone of copper deposits.

> **Hardness** 4.5
> **Density** 3.22
> **Luster** Vitreous
> **Cleavage** Perfect
> **Fracture** Uneven
> **Tenacity** Brittle

limonite

radial aggregates

thick tabular crystals

Similar minerals
Olivenite and libethenite have a different crystal form, as does pseudomalachite. Malachite and brochantite are almost always acicular. Occurrence with other phosphate minerals is characteristic.

Triclinic crystal form

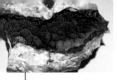

Rockbridgeite, Green Iron Ore
$(Fe,Mn)Fe_4[(OH)_5/(PO_4)_3]$

Occurrences In phosphate-bearing pegmatites as a conversion product of primary phosphate minerals and in phosphorus-rich brown iron deposits.

> *Hardness* 4.5
> *Density* 3.4
> *Luster* Vitreous
> *Cleavage* Present, but rarely recognizable
> *Fracture* Uneven
> *Tenacity* Brittle

Orthorhombic crystal form

Because of its green to black color and the radial reniform habit, rockbridgeite is also called green Glaskopf ("glass head"). Small tabular to acicular crystals are much rarer. In addition, rockbridgeite also occurs as fibrous, crusty, powdery and coarse.

radial aggregate

alternating color change

Similar minerals
Color and streak color are very characteristic. If the typical occurrence is taken into account, confusion is hardly possible; the closely related frondelite is always brown.

fibrous crusts

radial needle-like bundles

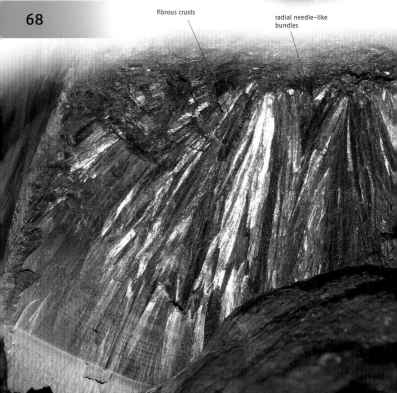

Dioptase

$Cu_6[Si_6O_{18}] \cdot 6\,H_2O$

Because of its intense emerald-green color, dioptase was formerly also called copper emerald because it was first thought to be emerald. It was first discovered in Altyn-Tyube in the steppe of Kazakhstan, where it was discovered by Cossacks. The long to short prismatic as well as acicular crystals are always raised and sometimes connected to radial aggregates. The best dioptase specimens in the world come from Africa: Sites in Zaire and the Tsumeb deposit in Namibia have provided rich material for mineral collectors around the world.

Occurrences In the oxidation zone of copper deposits, especially in siliceous rock.

> *Hardness* 5
> *Density* 3.3
> *Luster* Vitreous
> *Cleavage* Recognizable
> *Fracture* Conchoidal
> *Tenacity* Brittle

Trigonal crystal system

prismatic crystal

prismatic crystals

calcite

Similar minerals
Malachite has a different crystal form and effervesces with diluted hydrochloric acid; emerald is much harder and has a different crystal form.

Actinolite
(Ca,Fe)₂(Mg,Fe)₅[OH/Si₄O₁₁]₂

$(Ca,Fe)_2(Mg,Fe)_5[OH/Si_4O_{11}]_2$

crystal sheaves

Occurrences *Embedded in metamorphic rocks, especially talc and chlorite schists, in eclogites, in alpine-type fissures, in skarn rocks and contact metamorphic deposits.*
> **Hardness** 5.5–6
> **Density** 2.9–3.1
> **Luster** *Vitreous*
> **Cleavage** *Perfect, cleavage angle approx. 120°*
> **Fracture** *Uneven*
> **Tenacity** *Brittle*

Actinolite forms green prismatic columnar to acicular crystals. Radial aggregates are called actinolite; fine fibrous to capillary habits are referred to as amianthus or byssolite.

crystal column

amianthus

crystal column

talc schist

Monoclinic crystal form

Similar minerals
Pyroxenes have a different cleavage angle; tourmaline has a different crystal form and no cleavage; hornblende is more black.

Fassaite
Ca(Mg,Fe,Al)(Si,Al)₂O₆

$Ca(Mg,Fe,Al)(Si,Al)_2O_6$

prismatic crystals

Occurrences *In contact metamorphic rocks and volcanic ejecta.*

> **Hardness** 6
> **Density** 2.9–3.3
> **Luster** *Vitreous*
> **Cleavage** *Good*
> **Fracture** *Uneven*
> **Tenacity** *Brittle*

Fassaite seldom forms prismatic to tabular crystals, rather it mainly forms radial aggregates; usually coarse and granular.

chlorite

tabular crystal

Monoclinic crystal form

Similar minerals
Grossular and vesuvianite have a white streak color; actinolite has a different cleavage angle of 120°; diopside has a different streak color; olivine has a different crystal form.

uneven fracture

Aegirine, Acmite
$NaFeSi_2O_6$

Aegirine forms dark-green to black tabular to prismatic crystals. They are acicular, raised, and often embedded. In addition, there are also parallel-fibered aggregates or sun-shaped radial, spherical aggregates. The name aegirine was given to the mineral after the Nordic sea god Aegir, as it was found in Norway along the sea coast. The second name acmite indicates the often acicular formation of the mineral.

> **Occurrences** In alkali rocks and their pegmatites.

> > **Hardness** 5–6
> > **Density** 3.5–3.6
> > **Luster** Vitreous to greasy
> > **Cleavage** Perfect, cleavage angle approx. 90°
> > **Fracture** Uneven
> > **Tenacity** Brittle

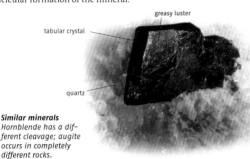

greasy luster

tabular crystal

quartz

Similar minerals
Hornblende has a different cleavage; augite occurs in completely different rocks.

Monoclinic crystal form

Hedenbergite
$CaFe[Si_2O_6]$

prismatic crystals

cleavage

Hedenbergite sometimes forms prismatic to tabular crystals, but usually forms radial aggregates, sometimes it is also coarse and granular. It is a rock-forming component of skarn rocks, e.g., in the iron deposits of the Italian island of Elba.

> **Occurrences** In contact metasomatic iron deposits, in volcanic ejecta, in sanidinites.

> > **Hardness** 6
> > **Density** 3.55
> > **Luster** Vitreous
> > **Cleavage** Recognizable, cleavage angle approx. 90°
> > **Fracture** Splintery
> > **Tenacity** Brittle

Similar minerals
The paragenesis makes hedenbergite unmistakable; ilvaite, which often occurs together with it, is black; actinolite has a different cleavage.

Monoclinic crystal form

radial fibers

Augite
(Ca,Mg,Fe)₂[(Si,Al)₂O₆]

Occurrences *In volcanic rocks as an accessory mineral, well-formed crystals especially in volcanic tuffs.*

> **Hardness** 6
> **Density** 3.3–3.5
> **Luster** Vitreous
> **Cleavage** Discernible prismatic cleavage, angle between the cleavage surfaces (cleavage angle) approx. 90°
> **Fracture** Conchoidal
> **Tenacity** Brittle

Monoclinic crystal form

Augite often forms well-formed short to long prismatic crystals, but can also be acicular, especially when grown in cavities in basaltic rocks. As an accessory mineral, it is granular to coarse. Fully formed crystals are embedded in volcanic tuffs in the Eifel or on the Canary Island of Lanzarote.

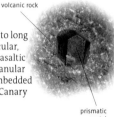

volcanic rock

prismatic crystal

embedded crystal

basalt

Similar minerals
Hornblende has a different cleavage and has more of a six-sided cross-section in contrast to the more four- or eight-sided cross-section of augite.

Gadolinite
Y₂FeBe₂[O/SiO₄]₂

Occurrences *In pegmatites, in alpine-type fissures.*

> **Hardness** 6.5
> **Density** 4–4.7
> **Luster** Pitchy to vitreous
> **Cleavage** Mostly unrecognizable
> **Fracture** Conchoidal
> **Tenacity** Brittle

Monoclinic crystal form

In pegmatites, gadolinite forms black embedded crystals and coarse masses (opaque, pitchy luster); in alpine-type fissures it forms raised green prismatic crystals (transparent, vitreous luster).

vitreous luster

prismatic crystal

albite

thick tabular crystal

pitchy luster

Similar minerals
The green, raised crystals of gadolinite are unmistakable; the black embedded crystals differ from other black minerals by their green streak color.

feldspar

Graphite

C

tabular crystal

calcite

Graphite forms dark-gray tabular, embedded crystals, flaky and scaly aggregates and coarse dense masses. Graphite rubs off on paper and, therefore, has been used since the 16th century, for example to make pencils. The hardness is determined by the amount of clay added and subsequent firing. In addition, graphite serves as a lubricant and is used in nuclear reactor construction.

Occurrences *In crystal-line schists, marbles, pegmatites.*

> ***Hardness*** *1*
> ***Density*** *2.1–2.3*
> ***Luster*** *Metallic to dull*
> ***Cleavage*** *Perfect basal*
> ***Fracture*** *Flaky*
> ***Tenacity*** *Flexible, brittle*

flaky aggregate

Similar minerals
Molybdenite is harder, its streak is slightly greenish when rubbed on a streak plate, graphite is more brownish; hematite has a red streak and is harder and brittle like ilmenite.

Hexagonal crystal form

metallic luster

Molybdenite

MoS₂

Molybdenite forms lead-gray tabular crystals, embedded sheets, as well as flaky and scaly aggregates. It has a dirty greenish color when rubbed on a streak plate. Artificial molybdenite serves as a particularly effective lubricant.

Occurrences *In peg-matites, in alpine-type fissures.*

> ***Hardness*** *1–1.5*
> ***Density*** *4.7–4.8*
> ***Luster*** *Metallic*
> ***Cleavage*** *Perfect basal*
> ***Fracture*** *Flaky*
> ***Tenacity*** *Flexible, brittle*

metallic luster

tabular crystals

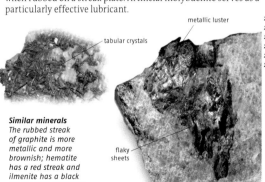

Similar minerals
The rubbed streak of graphite is more metallic and more brownish; hematite has a red streak and ilmenite has a black streak, both are much harder and brittle.

flaky sheets

quartz

Hexagonal crystal form

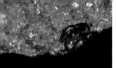

Covellite
CuS

Occurrences In
hydrothermal veins,
as a coating on other
sulfides, formed during
the weathering of pri-
mary copper sulfides.

> **Hardness** 1.5–2
> **Luster** Metallic to dull
> **Cleavage** Perfect basal
> **Fracture** Flaky
> **Tenacity** Brittle, thin
> flexible sheets

Covellite forms blue-black to
purple-black thin tabular crystals
and flakes, but usually coarsely
earthy coatings, especially on
pyrite. Its color becomes purple
when moistened with water.

thick sheets

tarnish

metallic luster

tabular crystals

Hexagonal crystal form

Similar minerals
*The blue–black
color and the
color change when
moistened prevent
confusion with other
minerals.*

tarnish

74

Nagyágite
Au(Pb,Sb,Fe)$_8$(Te,S)$_{11}$

Occurrences In hy-
drothermal and, in
particular, antimonite
quartz veins; more
rarely among other
sulfides in gold, silver,
and lead ore veins;
rarely metasomatically
in limestones.

> **Hardness** 1–1.5
> **Density** 7.4–7.6
> **Luster** Metallic
> **Cleavage** Perfect basal
> **Fracture** Hackly
> **Tenacity** Flexible sheets

Nagyágite sometimes forms lead-gray pseudotetragonal and
tabular crystals, but mostly flaky aggregates
and embedded masses. It was named
after where it was discovered: Nagyág
mine (today Săcărâmb) in Romania.

tabular crystal

quartz   flaky crystals

dolomite

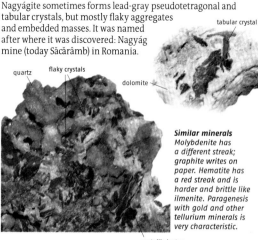

Orthorhombic
crystal form

Similar minerals
*Molybdenite has
a different streak;
graphite writes on
paper. Hematite has
a red streak and is
harder and brittle like
ilmenite. Paragenesis
with gold and other
tellurium minerals is
very characteristic.*

metallic luster

Argentite, Acanthite

Ag₂S

Argentite is isometric above 179°C and monoclinic below (acanthite). Argentite forms octahedral and cubic lead-gray crystals, while acanthite formed at low temperatures shows acicular crystals. Cubic crystals always consist of monoclinic acanthite, although they were formed at higher temperatures than cubic argentite. In the past, argentite was one of the most important silver ores.

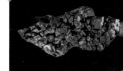

acicular

acanthite crystal

Occurrences *In hy-drothermal silver-ore veins, in the cementa-tion zone.*

> **Hardness** 2
> **Density** 7.3
> **Luster** Metallic, tarnish may cause it to become dull
> **Cleavage** Mostly indis-cernible
> **Fracture** Conchoidal
> **Tenacity** Malleable, sectile

Similar minerals
Galena is not mal-leable and cannot be cut; stephanite has a different crys-tal form, the same applies to hessite. Native silver is a lighter silvery color and more elastic.

Isometric crystal form

coarse masses

calcite

metallic luster

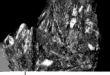

Antimonite, Stibnite
Sb_2S_3

Occurrences In hydrothermal and, in particular, antimonite quartz veins and in gold, silver, lead ore veins.

> **Hardness** 2
> **Density** 4.6–4.7
> **Luster** Metallic
> **Cleavage** Completely perfect
> **Fracture** Splintery, flaky
> **Tenacity** Flexible, brittle thin flakes and crystals

Antimonite forms prismatic to acicular crystals that are mostly raised, often bent, columnar and radial aggregates, but they are often also granular, coarse, dense. Antimonite often transforms into antimony oxides, e.g. cervantite or stibiconite, while retaining its crystal form (pseudomorphoses). In ancient times, stibnite was used as a face powder — and was by no means harmless; Cleopatra used it to emphasize her eye shadow.

acicular crystals

Orthorhombic crystal form

76

vertically striated crystals

prismatic crystal

Similar minerals Bismuthinite is much heavier and more yellowish white; arsenopyrite is harder and distinctly brittle; galena has a different cleavability. The perfect cleavability and the often bent crystals are very characteristic.

Bismuth

Bi

Bismuth has a silver-white color with a tinge of red, but is often darker. Well-formed crystals are rare as bismuth is usually coarse; because of its perfect cleavage it is flaky, skeletal and mostly embedded. Accompanying minerals are often cobalt and nickel minerals. Bismuth is used in the glass industry, in medicine, in steel production and, albeit rarely, in reactor technology.

Occurrences In hydrothermal veins, especially of the bismuth-cobalt-nickel formation; in pegmatites.

> *Hardness* 2–2.5
> *Density* 9.7–9.8
> *Luster* Metallic
> *Cleavage* Perfect, cleavage surfaces often striated
> *Fracture* Hackly to uneven
> *Tenacity* Brittle, but sectile

Trigonal crystal system

trigonal crystal

cleavage

metallic luster

77

metallic luster

dolomite

Similar minerals
The low hardness, color and striations on the cleavage surfaces make bismuth unmistakable.

Stephanite

Ag_5SbS_4

Occurrences In hydro-thermal silver ore veins in the cementation zone.

> **Hardness** 2.5
> **Density** 6.2–6.3
> **Luster** Metallic, dull from tarnish
> **Cleavage** Hardly recognizable
> **Fracture** Conchoidal to uneven
> **Tenacity** Brittle

Stephanite shows crystals appearing on six sides through twin formation. These are prismatic, thickly tabular, fused into rosette-shaped aggregates; less often stephanite is coarse. Stephanite used to be an important silver ore; it was named after the Austrian Archduke Stephan Franz Viktor.

tabular crystal

Orthorhombic crystal form

Similar minerals
Pyrargyrite is redder; argentite and hessite have a different crystal form; galena has perfect cleavage.

prismatic crystal

metallic luster

calcite

Hessite

Ag₂Te

Hessite forms pseudocubic, prismatic, often elongated, raised crystals; more rarely, it is fine-grained coarse.

long–needle crystal

quartz crystals

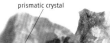

prismatic crystal

metallic luster

Occurrences In hydro-thermal silver and gold deposits, in subvolcanic deposits.

> ***Hardness*** 2.5
> ***Density*** 8.2–8.4
> ***Luster*** Metallic
> ***Cleavage*** None recognizable
> ***Fracture*** Uneven
> ***Tenacity*** Sectile, brittle

Monoclinic crystal form

Similar minerals
Argentite is difficult to distinguish from the much rarer Hessite, although it is somewhat harder; galena has a clearly visible cleavage; stephanite has a different crystal form; miargyrite has a different streak color.

79

Cylindrite

Pb₃Sn₄Sb₂S₁₄

small tubes

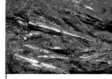

Cylindrite does not form actual crystals, instead it forms roll-like structures with a radial structure, which are usually embedded.

cross–section

radial tubes

Occurrences In subvolcanic tin ore deposits.

> ***Hardness*** 2.5
> ***Density*** 5.4
> ***Luster*** Metallic
> ***Cleavage*** None
> ***Fracture*** Conchoidal to uneven
> ***Tenacity*** Brittle

Similar minerals
The characteristic formation of cylindrite tubes does not allow for confusion; argentite is not brittle; galena has perfect cubic cleavage.

sphalerite

metallic luster

Galena
PbS

Occurrences In pegma-
tites, in hydrothermal
veins, as replace-
ment in limestones,
in sedimentary and
metamorphic sulfide
deposits.

> **Hardness** 2.5–3
> **Density** 7.2–7.6
> **Luster** Strong metallic
luster, often matte or blue
tarnish
> **Cleavage** Perfect cubic
> **Fracture** Splintery
> **Tenacity** Brittle

Galena forms raised crystals, mostly cubes, octahedrons or
combinations of both, as well as skeletal crystals, dendritic
formations and massive splintery masses. It is embedded and
intergrown with other sulfides. Twins can be flattened or
resemble six-sided plates. Galena is the most important lead ore
and, due to its often low (up to 1 weight percent) silver content,
also a common source of silver ore. The combination of lead and
silver mining makes the deposit worth exploiting.

Isometric crystal form

Similar minerals
With regard to
color, luster and
perfect cleavability,
galena can hardly be
confused; argentite
is much softer and
sectile; stephanite and
hessite have barely
recognizable cleav-
ability; arsenopyrite is
much harder.

quartz

cubes

chalcopyrite

cuboctahedral

octahedral
surface

metallic luster

Chalcocite

Cu₂S

Chalcocite is monoclinic below 103°C, hexagonal above. It forms tabular to prismatic twins with orthorhombic symmetry, as well as pseudohexagonal triplets, raised crystals, and is often coarse. In the past, chalcocite was an important ore for copper production, e.g. in Cornwall.

twin

Occurrences *In hydrothermal veins, especially in the cementation zone.*

> **Hardness** 2.5–3
> **Density** 5.7–5.8
> **Luster** Metallic, often dull from tarnish
> **Cleavage** Not visible
> **Fracture** Conchoidal
> **Tenacity** Brittle

Monoclinic pseudohexagonal crystal form

tabular crystal

81

Similar minerals
Its tenacity distinguishes chalcosite from other copper sulfides; digenite is somewhat bluer, but often difficult to distinguish.

metallic luster

tarnish

Bournonite, Cog Wheel Ore
PbCuSbS₃

Occurrences *In hydro-thermal lead and antimony veins.*

> **Hardness** 2.5–3
> **Density** 5.7–5.9
> **Luster** *Metallic, often dull from tarnish*
> **Cleavage** *Hardly visible*
> **Fracture** *Conchoidal*
> **Tenacity** *Brittle*

Orthorhombic crystal form

Bournonite forms thick tabular to prismatic crystals, often twins, reminiscent of cog wheels (hence the name cog wheel ore) and often coarse. Bournonite is rather rare, it is only used for lead extraction in conjunction with other lead ores.

Similar minerals *Tetrahedrite has a different crystal form, but is not easy to distinguish in rough aggregates from bournonite; galena has an excellent cleavability in contrast to bournonite; stephanite and hessite have a different crystal form.*

metallic luster

siderite

tabular crystal

striations

metallic luster

Bornite, Purple Copper Ore
Cu₅FeS₄

Bornite over 228°C is isometric, below is trigonal-pseudocubic. It rarely forms cubic or octahedral crystals, rather, it is more often coarse and embedded. Its color is reddish silver-gray in a fresh break, but it takes on a colorful tarnish within hours. Often, colorfully tarnished (sometimes even artificially colored) chalcopyrite is given out for genuine bornite.

tarnish

Occurrences *In pegmatites; hydrothermal ore veins, especially in the cementation zone; in alpine-type fissures.*

> **Hardness** 3
> **Density** 4.9–5.3
> **Luster** Metallic
> **Cleavage** Hardly visible
> **Fracture** Conchoidal
> **Tenacity** Brittle

Similar minerals
The typical tarnish colors distinguish boronite from almost all other sulfides; colorfully tarnished chalcopyrite is always yellow in a fresh fracture; chalcocite shows a different crystal form.

Isometric crystal form

octahedral crystal

quartz

calcite

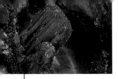

Dyscrasite
Ag₃Sb

Occurrences *In hydro-thermal silver ore deposits, especially in the cementation zone.*

> ***Hardness*** 3.5
> ***Density*** 9.4–10
> ***Luster*** Metallic
> ***Cleavage*** Mostly poorly recognizable
> ***Fracture*** Hackly
> ***Tenacity*** Brittle

Orthorhombic crystal form

The silver-gray dyscrasite forms prismatic, vertically striated crystals, which are usually embedded and poorly formed. Very typical are v-shaped twins and dendritic aggregates. Dyscrasite is often coarsely embedded.

fracture surface

calcite

Similar minerals
Argentite is softer; silver does tarnish as fast; the crystal form is completely different in each case. The v-shaped twins are very typical of dyscrasite.

re-entrant angle

twins

long tabular crystals

metallic luster

native arsenic

Tennantite

$Cu_3AsS_{3,25}$

The steel-gray to silver-gray tennantite forms tetrahedral crystals. These crystals are sometimes spherical due to their abundance of faces; this rounded variety of tennantite is called binnite after their occurrence in the Binn Valley (Switzerland). Raised crystals are rather rare; much more frequent are granular and coarse masses.

Occurrences In hydro-thermal veins, in subvolcanic deposits, in contact metasomatic deposits.

> **Hardness** 3–4
> **Density** 4.6–5.2
> **Luster** Metallic, often dull
> **Cleavage** None
> **Fracture** Conchoidal
> **Tenacity** Brittle

Similar minerals
Arsenopyrite is harder; galena has excellent cleavage properties; tetrahedrite is slightly lighter and has no reddish streak when rubbed but is difficult to distinguish from tennantite by simple means; enargite has perfect cleavage properties.

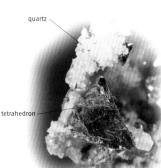

quartz

tetrahedron

Isometric crystal form

cubic crystals

barite

dull tarnish

Tetrahedrite

$Cu_3SbS_{3,25}$

Occurrences Seldom found in pegmatites, mostly in hydrothermal veins.

> **Hardness** 3–4
> **Density** 4.6–5.2
> **Luster** Metallic, often dull
> **Cleavage** None
> **Fracture** Conchoidal
> **Tenacity** Brittle

The steel-gray to silver-gray tetrahedrite usually only forms tetrahedra, and rarely forms crystals that are rich in faces. It often forms coarse aggregates or small vein fillings. Its name refers to its typical crystal form of tetrahedra.

chalcopyrite coating

Isometric crystal form

Similar minerals
Sphalerite and galena differ from tetrahedrite in their cleavability; chalcopyrite has a different color; tennantite has a slightly reddish streak when rubbed but is difficult to distinguish from tetrahedrite by simple means.

metallic luster

tetrahedron

siderite

Millerite

NiS

Millerite forms brass yellow to golden yellow metallic shiny acicular crystals that are often capillary or intergrown in bundles or sheaves. It can also form radial aggregates but very rarely is millerite coarse.

Occurrences In nickel deposits, where other nickel ores formed; in coal deposits as druses in the surrounding rock.

> - **Hardness** 3.5
> - **Density** 5.3
> - **Luster** Metallic
> - **Cleavage** Perfect, but almost never recognizable because of the acicular habit
> - **Fracture** Uneven
> - **Tenacity** Brittle

prismatic crystal

Trigonal crystal system

Similar minerals
Millerite's typical acicular to capillary habit and brass yellow color rule out confusion. The rare acicular pyrite is much harder.

acicular crystals

calcite

radial appearance

metallic luster

Enargite
Cu₃AsS₄

Occurrences *In arsenic-rich copper ore veins, in subvolcanic deposits.*

> **Hardness** 3.5
> **Density** 4.4
> **Luster** Metallic
> **Cleavage** Perfect prismatic
> **Fracture** Uneven
> **Tenacity** Brittle

Enargite is steel gray to iron black with a purple hue. Its crystals are often pseudohexagonal, prismatic and often vertically striated. You can also find star-shaped triplets, next to which enargite forms radial aggregates and, most often, coarse masses, which always show perfect cleavage.

Orthorhombic crystal form

coarse enargite

pyrite

tabular crystals

metallic luster

Similar minerals
Arsenopyrite is harder; tetrahedrite has a different crystal form and no cleavability. Chalcocite isn't brittle; bournonite has a different crystal form.

Chalcopyrite

CuFeS₂

The crystals of chalcopyrite are brass yellow with a greenish hue and often tarnished. The crystals are similar to tetrahedra and octahedron, often formed as twins, but most of the chalcopyrite is coarse. Chalcopyrite is the most important copper ore in the world.

Occurrences In granites and gabbros, in pegmatites and tin ore veins, in hydrothermal veins and black schists.

> **Hardness** 3–4
> **Density** 4.2–4.3
> **Luster** Metallic
> **Cleavage** Hardly recognizable
> **Fracture** Conchoidal
> **Tenacity** Brittle

octahedron-like crystal

metallic luster

barite

Tetragonal crystal form

colorful tarnish

Similar minerals
Pyrite is much harder; pyrrhotite is a more brown color; gold is softer and is sectile and malleable. Yellow tarnished bismuth shows its true color when scratched.

tetrahedron-like crystals

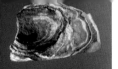

Arsenic, native
As

Occurrences *Silver and cobalt ore veins bearing arsenic, hydro-thermal veins.*

> **Hardness** *3–4*
> **Density** *7.06*
> **Luster** *Fresh surfaces have a metallic luster; quickly turns dark and dull after only a few hours*
> **Cleavage** *Not visible*
> **Fracture** *Uneven*
> **Tenacity** *Brittle*

Trigonal crystal system

Black-gray to black native arsenic sometimes forms cube-like to acicular crystals. Usually it is shell-like, spherical, botryoidal, radial, dense.

dull tarnish

shell-like structure

reniform aggregates

Similar minerals
Reniform pyrite and marcasite are much harder; goethite has a brown streak color; galena has excellent cubic cleavage; dyscrasite has a different crystal form and is not brittle.

Iron (Native)
Fe

Occurrences *Embedded in terrestrial basalts, as a component of mete-orites, iron meteorites consist almost entirely of nickel-containing iron.*

> **Hardness** *4–5*
> **Density** *7.88*
> **Luster** *Metallic*
> **Cleavage** *None*
> **Fracture** *Hackly*
> **Tenacity** *Ductile*

Isometric crystal form

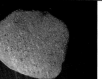

Steel-gray shiny iron forms small, embedded scales, drops, irregular masses — no well-formed crystals and it is never raised. Native iron is very rare and plays no role in industrial use. Among collectors, however, it is precisely for this reason that it is in great demand.

basalt

inclusion of iron

basalt

metallic luster

polished slab

Similar minerals
The paragenesis in basalts and the tenacity of iron prevent mix-ups; magnetite is harder.

Pyrrhotite
FeS

Pyrrhotite is bronze with a brownish streak. It forms, albeit rarely, prismatic, thick to thin tabular crystals, which sometimes grow together to form rosettes. Most of the time, however, it is crudely ingrown. Pyrrhotite is also magnetic.

prismatic crystals

Occurrences *In hydrothermal veins and metamorphic gravel deposits, in gold quartz veins and subvolcanic deposits.*

> ***Hardness*** 4
> ***Density*** 4.6
> ***Luster*** *Metallic*
> ***Cleavage*** *Rarely visible*
> ***Fracture*** *Uneven*
> ***Tenacity*** *Brittle*

Similar minerals
Pyrite and chalcopyrite are much more yellow, pyrite is also harder; sphalerite has a per-fect cleavage.

calcite

Hexagonal crystal form

metallic luster

tabular crystal

quartz crystals

Platinum (Native)
Pt

Occurrences In placers and quartz veins, embedded in basic to ultrabasic rocks.

> **Hardness** 4–4.5
> **Density** 21.4
> **Luster** Metallic
> **Cleavage** None
> **Fracture** Hackly
> **Tenacity** Ductile, malleable (not quite as malleable as gold)

Native platinum sometimes forms cubic crystals; more often it forms rounded nuggets, usually weighing only a few grams, but they can also weigh up to several kilograms. Platinum often also forms small plates or grains, mostly loose and rarely embedded. Platinum belongs to the precious metals family but is rarely used in the jewelry industry. It is particularly important in the chemical industry because of its temperature and chemical resistance. Much of the platinum consumed today is contained in automotive catalytic converters.

Isometric crystal form

granular aggregate

platinum nugget

metallic luster

Similar minerals
Silver is softer; iron is magnetic; sperrylite is not as malleable and is much harder.

Ilmenite, Titanic Iron Ore
FeTiO₃

Ilmenite forms black rhombohedral, thick to thin tabular crystals, sometimes growing together to form rosettes (ilmenite roses) and is often granular, coarse. Today, ilmenite is the most significant ore for the lightweight metal titanium.

thick tabular crystal

isometric crystal

feldspar

Occurrences Embedded in igneous rocks, pegmatites, rounded pebbles in placers, tabular crystals and ilmenite roses in alpine-type fissures.

> **Hardness** 5–6
> **Density** 4.5–5
> **Luster** Metallic, often dull from tarnish
> **Cleavage** None
> **Fracture** Conchoidal to uneven
> **Tenacity** Brittle

Similar minerals
Magnetite has a different crystal form (mostly octahedron); hematite has a red streak.

Trigonal crystal system

Arsenopyrite
FeAsS

Arsenopyrite forms tin-white to steel-gray tabular to prismatic, sometimes octahedron-like, crystals that are often embedded and coarse masses. Arsenopyrite was unpopular in ore mining because it formed toxic residues during smelting.

crystal striation

thick tabular crystal

siderite

regular association

Occurrences In pegmatites and tin ore veins, but especially in hydrothermal veins.

> **Hardness** 5.5–6
> **Density** 5.9–6.2
> **Luster** Metallic
> **Cleavage** Indiscernible
> **Fracture** Uneven
> **Tenacity** Brittle

Orthorhombic crystal form

Similar minerals
Pyrite and marcasite have a golden yellow color; pyrrhotite is somewhat softer and more brownish.

Columbite
(Fe,Mn)(Nb,Ta)$_2$O$_6$

Columbite is the name of a mineral group. This
group contains the mixture members of the
minerals niobite (Fe,Mn)Nb$_2$O$_6$ and
tantalite (Fe,Mn)Ta$_2$O$_6$. If the exact
ratio of niobium to tantalum is not
known, these minerals are
simply called columbite.
Columbites form black tabular
to acicular crystals, which are
mostly embedded. The
manganese-richest members
(manganotantalite) are
brown-red to red.

manganotantalite

conchoidal
fracture

Orthorhombic crystal form

Similar minerals
Hematite has a dif-
ferent streak; ilmenite
has a different crystal
form; niobite and
tantalite are indistin-
guishable by simple
means.

tabular
crystal

pitchy luster

vertical striations

Ilvaite
$CaFe_2^{2+}Fe^{3+}[OH/O/Si_2O_7]$

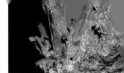

The black ilvaite forms prismatic crystals, radial to columnar aggregates, and coarse and granular masses.

prismatic crystals

dull surface

regular association

> **Occurrences** In iron-rich contact deposits.

> **Hardness** 5.5–6
> **Density** 4.1
> **Luster** Vitreous, somewhat resinous, often dull
> **Cleavage** Hardly recognizable
> **Fracture** Conchoidal
> **Tenacity** Brittle

Similar minerals
Tourmaline is harder; actinolite has a different paragenesis and the typical cleavage properties of the members of the amphibol group (cleavage surfaces at an angle of about 120°).

Orthorhombic crystal form

Pyrolusite, Manganese Dioxide
MnO_2

Pyrolusite forms silver-gray prismatic to thick tabular crystals, radial aggregates, and, often, dull black earthy, crusty coatings.

metallic luster

acicular crystals

thick tabular crystals

> **Occurrences** In hydrothermal veins, in the oxidation zone, in sediments as small granules (ooliths).

> **Hardness** 6 (but often lower in aggregates)
> **Density** 4.9–5.1
> **Luster** Metallic to dull
> **Cleavage** None
> **Fracture** Conchoidal, crumbly to fibrous in aggregates
> **Tenacity** Brittle

Similar minerals
Manganite has a brown streak; antimonite is not as brittle and much softer; romanechite is somewhat harder.

Tetragonal crystal form

Pyrite, Fool's Gold
FeS₂

Occurrences *In rocks
of all kinds, in meta-
morphic deposits, in
hydrothermal veins.*

> | ***Hardness*** 6–6.5
> | ***Density*** 5–5.2
> | ***Luster*** Metallic
> | ***Cleavage*** None
> | ***Fracture*** Conchoidal
> | ***Tenacity*** Brittle

Pyrite forms brass yellow crystals: Cubes with striated surfaces,
octahedron, pentagonal dodecahedron, sometimes extremely
rich in crystal faces, raised and embedded crystals. The crystals
are often up to many centimeters in size and up to several tons in
weight. In the past, pyrite was even processed into polished
stones because of its golden color. Pyrite also forms radial and
reniform aggregates, rarely spheres or
disc-shaped aggregates (pyrite suns),
and it is often coarse. Due to its
metallic sheen and golden color,
pyrite was also called fool's
gold.

Isometric crystal form

pentagonal dodecahedron

quartz

spherical crystal aggregate

dolomite crystals

metallic luster

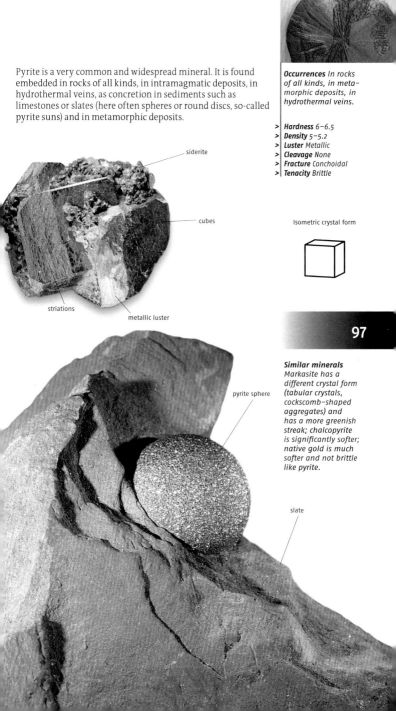

Pyrite is a very common and widespread mineral. It is found embedded in rocks of all kinds, in intramagmatic deposits, in hydrothermal veins, as concretion in sediments such as limestones or slates (here often spheres or round discs, so-called pyrite suns) and in metamorphic deposits.

siderite

cubes

striations

metallic luster

Occurrences In rocks of all kinds, in metamorphic deposits, in hydrothermal veins.

> *Hardness* 6–6.5
> *Density* 5–5.2
> *Luster* Metallic
> *Cleavage* None
> *Fracture* Conchoidal
> *Tenacity* Brittle

Isometric crystal form

97

Similar minerals
Markasite has a different crystal form (tabular crystals, cockscomb-shaped aggregates) and has a more greenish streak; chalcopyrite is significantly softer; native gold is much softer and not brittle like pyrite.

pyrite sphere

slate

Chloanthite

(Ni,Co)As₂₋₃

(Ni,Co)As$_{2-3}$

Occurrences *In hydro-thermal deposits, in cobalt–nickel deposits.*

> **Hardness** *5.5–6*
> **Density** *6.5*
> **Luster** *Metallic*
> **Cleavage** *Poor*
> **Fracture** *Conchoidal*
> **Tenacity** *Brittle*

Chloanthite forms silver-gray cubic and octahedral crystals, also cubes in combination with the rhombic dodecahedron, raised and embedded, coarse. Chloanthite is an important nickel ore. The metal nickel serves, for example, as an alloy metal in the production of particularly resistant steels.

cube surface

rhombic dodecahedron surface

spherical crystal aggregates

Isometric crystal form

Similar minerals
Pyrite is more yellow and harder; arseno-pyrite has a different crystal form, galena has perfect cubic cleavage.

metallic luster

98

Marcasite, White Iron Pyrite, Cockscomb Pyrite

FeS₂

Occurrences *In hydro-thermal, low-tem-perature displacement deposits, as concretions, but also in the form of crystals in sediments, especially limestones and marls.*

> **Hardness** *6–6.5*
> **Density** *4.8–4.9*
> **Luster** *Metallic*
> **Cleavage** *Poor*
> **Fracture** *Uneven*
> **Tenacity** *Brittle*

Brass yellow tabular crystals are often intergrown into serrated, comb-shaped groups. Frequently radial and shell-like, reniform aggregates are spheres. Marcasite is particularly sensitive to moisture, which is why it decays quickly. It must, therefore, be kept absolutely dry.

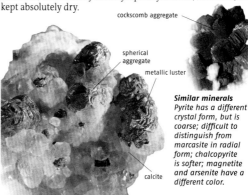

cockscomb aggregate

spherical aggregate

metallic luster

Orthorhombic crystal form

Similar minerals
Pyrite has a different crystal form, but is coarse; difficult to distinguish from marcasite in radial form; chalcopyrite is softer; magnetite and arsenite have a different color.

calcite

Magnetite, Lodestone
Fe_3O_4

Black magnetite forms octahedrons, more rarely rhombic dodecahedrons and cubes that are raised and embedded. Usually, however, it forms coarse masses. Magnetite is highly magnetic; it can attract smaller iron particles such as iron filings, nails and paper clips. It is a very important iron ore.

Similar minerals
All similar minerals are not, or only weakly, magnetic; chromite has a light brown streak.

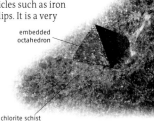

embedded octahedron

chlorite schist

Occurrences *Embedded in igneous rocks, in displacement and metamorphic deposits, crystals in chlorite and talc shale, in hydro-thermal veins.*

> **Hardness** 6–6.5
> **Density** 5.2
> **Luster** Dull metallic
> **Cleavage** Hardly recognizable
> **Fracture** Conchoidal
> **Tenacity** Brittle

Isometric crystal form

adularia

raised octahedron

metallic luster

Epidote
$Ca_2(Fe,Al)Al_2[O/OH/SiO_4/Si_2O_7]$

Occurrences In druses and cavities of pegmatites, in epidote schists, in fissures in granites and metamorphic rocks.

> **Hardness** 6–7
> **Density** 3.3–3.5
> **Luster** Vitreous
> **Cleavage** Poorly visible
> **Fracture** Conchoidal
> **Tenacity** Brittle

Epidote forms prismatic, rarely thick tabular crystals, which can vary in color from yellow-green to black-green and almost black. Similarly, its streak color varies from greenish to greenish black. Epidote also forms radial aggregates; it is often coarse and embedded in rocks.

translucent

regular association of three crystals

Monoclinic crystal form

Similar minerals
In contrast to epidote, augite, hornblende and actinolite have perfect cleavage properties; tourmaline has a different crystal form.

100

regular association

prismatic crystal

amianthus

Romanechite

$BaMn_8O_{16}(OH)_4$

Romanechite very rarely forms crystals, mostly reniform, stalactitic aggregates or radial masses and is often earthy, coarse. If it occurs in larger quantities, romanechite can serve as manganese ore.

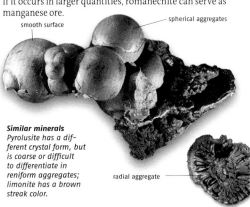

smooth surface

spherical aggregates

Occurrences In weathering deposits, as concretions in sediments, as displacements in limestones, as crusts on limonite.

> **Hardness** 6–6.5
> **Density** 6.3–6.45
> **Luster** Metallic to dull
> **Cleavage** None
> **Fracture** Uneven
> **Tenacity** Brittle

Similar minerals
Pyrolusite has a different crystal form, but is coarse or difficult to differentiate in reniform aggregates; limonite has a brown streak color.

radial aggregate

Monoclinic crystal form

Gahnite

$ZnAl_2O_4$

Gahnite usually forms octahedral crystals, partly with rhombic dodecahedron surfaces; they are raised and embedded. Seldom is gahnite coarse. The largest and best gahnite crystals come from the Silberberg near Bodenmais in the Bavarian Forest.

octahedron

quartz

schist

Occurrences Accessory mineral in granites and pegmatites, in metamorphic rocks, in metamorphic sulfide deposits.

> **Hardness** 7.5–8
> **Density** 4.5–4.8
> **Luster** Vitreous to greasy
> **Cleavage** Poor
> **Fracture** Conchoidal
> **Tenacity** Brittle

octahedron with rhombic dodecahedron

Isometric crystal form

Similar minerals
Magnetite is pure black and highly magnetic; spinel is much lighter; franklinite is softer.

Kaolinite

$Al_2Si_2O_5(OH)_4$

Occurrences *Forms from the weathering of silicates, particularly feldspar.*

> **Hardness** *1*
> **Density** *2.6*
> **Luster** *Dull (earthy)*
> **Cleavage** *Perfect*
> **Fracture** *Dull, crumbly*
> **Tenacity** *Brittle, sculptural*

Triclinic crystal form

Kaolinite crystals are microscopically small; to our eyes it is powdery, earthy. It forms white to gray or brown sculptural masses.

compacted chunks of kaolin

kaolinized feldspar

kaolin powder

Similar minerals
The low hardness and the sculptural characteristics make kaolinite quite unmistakable, however other clay minerals cannot be distinguished by simple means.

Vermiculite

$(Mg,Fe,Al)_3[(OH)_2/(Al,Si)_2Si_2O_{10}] \cdot 4\ H_2O$

Occurrences *Formed from the hydrothermal transformation of igneous rocks, especially from biotite.*

> **Hardness** *1*
> **Density** *2.4–2.7*
> **Luster** *Pearly to dull*
> **Cleavage** *Perfect basal*
> **Fracture** *Uneven to foliated*
> **Tenacity** *Flexible, brittle*

Monoclinic crystal form

Vermiculite forms white to brown or golden-yellow tabular crystals and platy, flaky, curved vermicular, coin roll-like aggregates of individual sheets. When heated (using a lighter!), it expands up to 50 times its original volume.

prismatic crystal

cleavage plane

platy cleavage

Similar minerals
Its low hardness and expansion when heated make vermiculite unmistakable.

Talc, Steatite, Soapstone
$Mg_3[(OH)_2/Si_4O_{10}]$

Talc very rarely develops well-formed crystals, usually it is foliated, dense with a reniform surface. It often gives the appearance of other minerals, e.g. quartz or dolomite crystals (pseudomorphism). It is white, yellowish, brownish to green. Soapstone is very easy to carve.

Occurrences Ingrown in metamorphic rocks, as a primary constituent of talc shale, as a filling in fissures in serpentines.

> *Hardness* 1
> *Density* 2.7–2.8
> *Luster* Pearly to greasy
> *Cleavage* Perfect basal
> *Fracture* Uneven to foliated
> *Tenacity* Flexible, brittle

Similar minerals
Talc's low hardness and greasy feel makes it quite distinctive.

soapstone in the form of a quartz crystal (pseudomorphism)

dense soapstone

Monoclinic crystal form

radial crystals

pearly luster

Halotrichite
FeAl₂[SO₄]₄ · 22 H₂O

$FeAl_2[SO_4]_4 \cdot 22\,H_2O$

Occurrences As efflorescence on rocks rich in aluminum, in old mines, often a weathering product of pyrite.

> **Hardness** 1.5
> **Density** 1.73–1.79
> **Luster** Silky to vitreous
> **Cleavage** None
> **Fracture** Fibrous
> **Tenacity** Brittle

Halotrichite is acicular and fibrous. It forms fibrous aggregates, with curved coils and is powdery, earthy. Its individual white to slightly yellowish fibers are flexible.

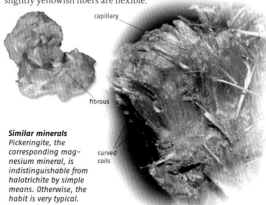

capillary

fibrous

Orthorhombic crystal form

Similar minerals
Pickeringite, the corresponding magnesium mineral, is indistinguishable from halotrichite by simple means. Otherwise, the habit is very typical.

curved coils

Aurichalcite, Mountain Brass
(Zn,Cu)₅[(OH)₃/CO₃]₂

$(Zn,Cu)_5[(OH)_3/CO_3]_2$

acicular crystals

Occurrences In the oxidation zone of copper-zinc deposits.

> **Hardness** 2
> **Density** 3.6–4.3
> **Luster** Silky to pearly
> **Cleavage** Perfect
> **Fracture** Flaky
> **Tenacity** Brittle

Aurichalcite forms platy, acicular, radial and aggregate cluster habits, crystal druse. The fibrous aggregates have a typical light blue color. It effervesces when dabbed with diluted hydrochloric acid.

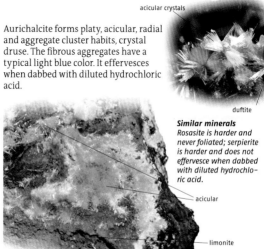

duftite

Similar minerals
Rosasite is harder and never foliated; serpierite is harder and does not effervesce when dabbed with diluted hydrochloric acid.

acicular

limonite

Chlorargyrite, Silver Chloride
AgCl

Chlorargyrite sometimes forms cubic or octahedral whitish to yellowish gray crystals. It mainly forms reniform, crusty coatings or partly large heavy, coarse masses. It can be cut with a knife.

reniform

greasy luster

limonite

Occurrences In the oxidation zone and cementation zone of silver deposits, especially in desert-like climates.

> *Hardness* 1.5
> *Density* 5.5–5.6
> *Luster* Adamantine to greasy
> *Cleavage* None
> *Fracture* Hackly
> *Tenacity* Malleable, sectile

Isometric crystal form

Similar minerals
Color, luster and tenacity are very characteristic. Limestone crystals effervesce when dabbed with hydrochloric acid.

105

Pyrophyllite, Agalmatolite
Al_2[(OH)_2/Si_4O_{10}]

Pyrophyllite forms platy, radial-beam aggregates, crystal druse, and fibrous aggregates. It is often reniform, coarse, dense. Its color is white, but also brownish, yellowish, greenish.

Occurrences In crystalline schist, on ore veins.

> *Hardness* 1.5
> *Density* 2.8
> *Luster* Pearly to greasy
> *Cleavage* Perfect basal
> *Fracture* Uneven
> *Tenacity* Flexible, brittle

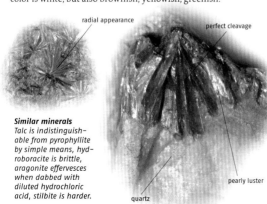

radial appearance

perfect cleavage

Similar minerals
Talc is indistinguishable from pyrophyllite by simple means, hydroboracite is brittle, aragonite effervesces when dabbed with diluted hydrochloric acid, stilbite is harder.

pearly luster

quartz

Monoclinic crystal form

Vivianite, Blue Iron Ore
$Fe_3[PO_4]_2 \cdot 8\,H_2O$

Occurrences In pegmatites, in the oxidation zone of ore deposits, in sediments.

> ***Hardness*** 2
> ***Density*** 2.6–2.7
> ***Luster*** Pearly
> ***Cleavage*** Perfect
> ***Fracture*** Flaky
> ***Tenacity*** Thin flexible sheets, brittle

Vivianite forms blue to greenish blue prismatic to tabular crystals and spherical, radial aggregates. It is coarse, powdery, earthy and forms crusts. It is seldom found as a petrifying medium, e.g., in fossilized cones.

typical oblique end face

parquet-like surface

single crystal

Monoclinic crystal form

Similar minerals
Azurite effervesces when dabbed with diluted hydrochloric acid, lazulite has a greasy luster and is harder, both lack the brittle tenacity of vivianite. Chalcanthite, unlike vivianite, is soluble in water.

106

limonite

radial appearance

sun-shaped aggregate

Annabergite

$Ni_3[AsO_4]_2 \cdot 8\,H_2O$

Annabergite rarely forms grass-green prismatic to tabular crystals, but they always have obliquely cut end faces. It also forms acicular aggregates, but is usually coarse, earthy and crusty. Green coatings on annabergite are a typical sign of nickel ores.

typical oblique end face

tabular crystal

crystal clusters

Occurrences In the oxidation zone of nickel deposits.

> **Hardness** 2
> **Density** 3–3.1
> **Luster** Vitreous
> **Cleavage** Completely perfect
> **Fracture** Flaky
> **Tenacity** Brittle, thin flexible sheets

Similar minerals
Malachite and other green copper minerals are darker – annabergite has a very distinct green; malachite effervesces when dabbed with diluted hydrochloric acid.

Monoclinic crystal form

Thomsenolite

$CaNaAlF_6 \cdot H_2O$

typical oblique end

The white to yellowish, often yellowish brown, coated crystals are long to short prisms, rarely tabular. They are often typically striated with one-sided oblique ends.

prismatic crystal

colored brown by iron oxides

crystal druse

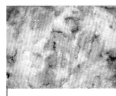

Occurrences As conversion product in druses from the alteration of cryolite in pegmatites, in druses in alkaline rocks.

> **Hardness** 2
> **Density** 2.98
> **Luster** Vitreous
> **Cleavage** Perfect
> **Fracture** Uneven
> **Tenacity** Brittle

Monoclinic crystal form

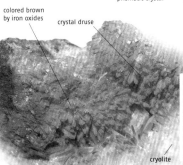

Similar minerals
Pachnolite crystals have a diamond-shaped cross-section and are not striated; ralstonite has a different crystal form and is harder.

cryolite

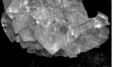

Sylvite
KCl

Occurrences In salt deposits, as efflorescence in steppes, rarely on volcanic rocks at the exit points of gases.

> **Hardness** 2
> **Density** 1.99
> **Luster** Vitreous
> **Cleavage** Perfect cubic
> **Fracture** Uneven
> **Tenacity** Brittle

Isometric crystal form

Sylvite forms cubic crystals, often with beveled edges. It is colorless, white or often orange to brownish and tastes bitter in contrast to common salt. Crystal formation is rather rare, granular coarse masses and layers are much more common.

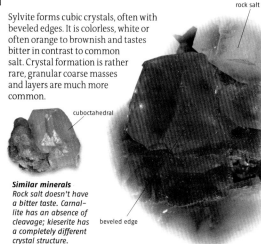

rock salt

cuboctahedral

beveled edge

Similar minerals
Rock salt doesn't have a bitter taste. Carnallite has an absence of cleavage; kieserite has a completely different crystal structure.

Borax
$Na_2[B_4O_5(OH)_4] \cdot 8H_2O$

Occurrences In borax lakes, especially in arid areas, in deserts.

> **Hardness** 2–2.5
> **Density** 1.7–1.8
> **Luster** Vitreous to dull
> **Cleavage** Sometimes visible
> **Fracture** Conchoidal
> **Tenacity** Brittle

Monoclinic crystal form

Borax is typically chalky white, rarely transparent, crystals that are long to short prisms, rarely tabular. Crusts that effloresce and powdery masses or coarse aggregates are common.

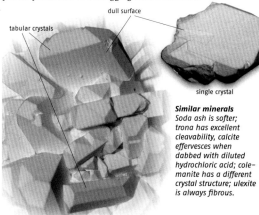

dull surface

tabular crystals

single crystal

Similar minerals
Soda ash is softer; trona has excellent cleavability, calcite effervesces when dabbed with diluted hydrochloric acid; colemanite has a different crystal structure; ulexite is always fibrous.

Sulfur

S

The yellow, raised sulfur crystals often form bipyramids, pointed pyramids, and rarely tabular crystals. Sulfur is often granular, fibrous, reniform, stalactitic, earthy, powdery. The crystals are especially sensitive to heat and crack in a warm hand.

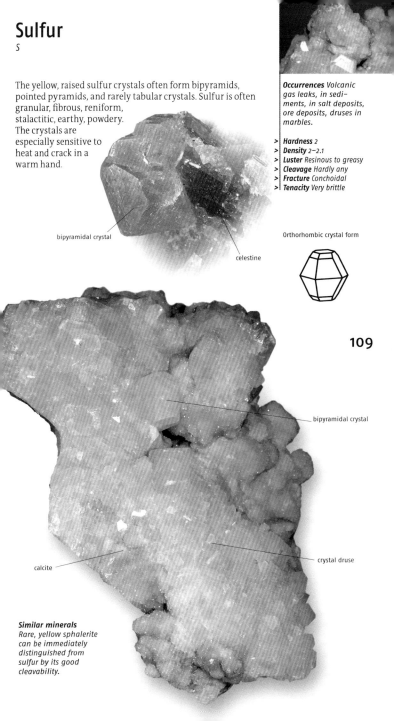

bipyramidal crystal

celestine

Occurrences Volcanic gas leaks, in sediments, in salt deposits, ore deposits, druses in marbles.

> *Hardness* 2
> *Density* 2–2.1
> *Luster* Resinous to greasy
> *Cleavage* Hardly any
> *Fracture* Conchoidal
> *Tenacity* Very brittle

Orthorhombic crystal form

bipyramidal crystal

crystal druse

calcite

Similar minerals
Rare, yellow sphalerite can be immediately distinguished from sulfur by its good cleavability.

Rock salt, Halite
NaCl

Occurrences *In sedimentary rocks, in salt deposits.*

> **Hardness** 2
> **Density** 2.1–2.2
> **Luster** Vitreous
> **Cleavage** Perfect cubic
> **Fracture** Conchoidal
> **Tenacity** Brittle

Rock salt is most often white, but can also be reddish, brownish, yellow and ink blue. It almost only forms cubes but can very rarely form octahedrons. Often it has raised crystals, but fibrous rock salt and coarse, granular masses also exist. Rock salt is water-soluble and typically tastes salty.

Isometric crystal form

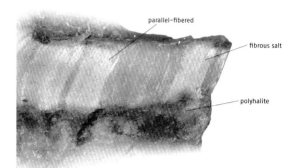

parallel–fibered

fibrous salt

polyhalite

skeletal habit

cubic crystals

Similar minerals
Fluorite is harder and not water-soluble, as is calcite, which also has a different crystal form.

Gypsum, Selenite
$CaSO_4 \cdot 2\,H_2O$

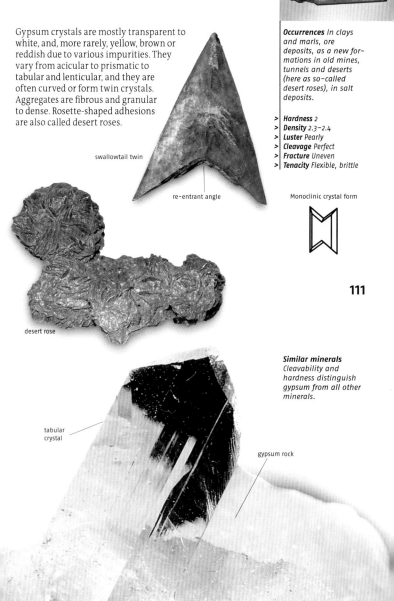

Gypsum crystals are mostly transparent to white, and, more rarely, yellow, brown or reddish due to various impurities. They vary from acicular to prismatic to tabular and lenticular, and they are often curved or form twin crystals. Aggregates are fibrous and granular to dense. Rosette-shaped adhesions are also called desert roses.

swallowtail twin

re-entrant angle

Occurrences In clays and marls, ore deposits, as a new for- mations in old mines, tunnels and deserts (here as so-called desert roses), in salt deposits.

> **Hardness** 2
> **Density** 2.3–2.4
> **Luster** Pearly
> **Cleavage** Perfect
> **Fracture** Uneven
> **Tenacity** Flexible, brittle

Monoclinic crystal form

111

desert rose

Similar minerals Cleavability and hardness distinguish gypsum from all other minerals.

tabular crystal

gypsum rock

Ettringite
$Ca_6Al_2(SO_4)_3(OH)_{12} \cdot 24\ H_2O$

Occurrences In volcanic rocks in calcium-rich inclusions, in meta-morphic manganese deposits.

> **Hardness** 2–2.5
> **Density** 1.77
> **Luster** Vitreous
> **Cleavage** Hardly recognizable
> **Fracture** Uneven
> **Tenacity** Brittle

Hexagonal crystal form

Ettringite crystals are mostly transparent to white, and, more rarely, are yellow or brown due to various impurities. They vary from acicular to prismatic to tabular. Aggregates are fibrous, radial and granular to dense.

radial appearance

Similar minerals
Calcite and afwillite have a different crystal form; calcite also has perfect cleavability and effervesces when dabbed with diluted hydrochloric acid.

prismatic crystal

colored yellow by iron

Kämmererite, Chromian Clinochlore

$Ca_6Al_2(Fe,Mg,Al,Cr)_6[(OH)_2/(Si,Al)_4O_{10}]$

The thick to thin tabular, often rhombohedron-like, crystals are intensely pink-purple in color. In addition, kämmererite forms platy or granular coatings of the same color on chrome ore.

Occurrences Conversion product of chromite in fissures of serpentine and crude chromite ore.

> **Hardness** 2
> **Density** 2.9–3.3
> **Luster** Vitreous, pearly on cleavage planes
> **Cleavage** Perfect basal
> **Fracture** Flaky
> **Tenacity** Brittle, flexible

Similar minerals
Mica is harder and elastic; clinochlore is green, the pinkish purple color in conjunction with the chrome-rich paragenesis makes kämmererite unmistakable.

Monoclinic crystal form

crystal striation

isometric crystal

colored red by chrome

113

Muscovite
KAl₂[(OH,F)₂/AlSi₃O₁₀]

$KAl_2[(OH,F)_2/AlSi_3O_{10}]$

Occurrences *Rock-forming component in granites, gneisses, mica schists, sandstones, marbles; they do not form in volcanic rocks; raised crystals in alpine-type fissures or in druses in pegmatites.*

> **Hardness** *2–2.5*
> **Density** *2.78–2.88*
> **Luster** *Pearly*
> **Cleavage** *Extremely perfect basal*
> **Fracture** *Flaky*
> **Tenacity** *Brittle, elastic sheets*

Monoclinic crystal form

Muscovite (white mica) forms silvery to light brown, or slightly greenish, thin to thick tabular six-sided crystals, as well as sheets, scales and rosette- or star-shaped aggregates. In most cases, it is an embedded rock-forming component, thus making it, for example, a primary constituent of mica schist. In pegmatites, it can form sheets up to several square meters.

crystal set

albite

114

Similar minerals
Talc and chlorite are softer, and their sheets are not elastic; biotite and phlogopite are almost always signifi-cantly darker.

tabular crystals

feldspar

pearly luster

Phlogopite

$KMg_3[(F,OH)_2/AlSi_3O_{10}]$

Phlogopite forms brown to blackish tabular, six-sided crystals. It is pseudohexagonal, sometimes prismatic, foliated, flaky, embedded and raised. Sheets can reach over one square meter in size.

six-sided crystal

calcite

cleavage plane

foliated

Occurrences *In marbles, metamorphic dolomites and pegmatites, in contact metasomatic formations, volcanic rock fissures.*

> ***Hardness*** 2–2.5
> ***Density*** 2.75–2.97
> ***Luster*** *Vitreous*
> ***Cleavage*** *Extremely perfect basal*
> ***Fracture*** *Flaky*
> ***Tenacity*** *Thin flexible sheets*

Similar minerals
Biotite occurs in other paragenesis and is always brown; whereas muscovite is always lighter (white) in color; clinochlore is much softer.

Monoclinic crystal form

115

Biotite

$K(Mg,Fe)_3[(OH)_2/(Al,Fe)Si_3O_{10}]$

Biotite forms black, six-sided crystals. They are tabular, rarely prismatic, and also form rosette-shaped aggregates, sheets or scales. During weathering, biotite becomes golden to a shiny metallic reddish color (rubellan).

volcanic tuff

quartz

rubellan

tabular crystal

cleavage plane

Occurrences *As an accessory mineral in igneous and metamorphic rocks, sometimes as raised crystals in fissures, in druses in volcanic ejecta.*

> ***Hardness*** 2.5–3
> ***Density*** 2.8–3.2
> ***Luster*** *Pearly*
> ***Cleavage*** *Extremely perfect basal*
> ***Fracture*** *Flaky*
> ***Tenacity*** *Brittle, elastic sheets*

Similar minerals
Chlorite and talc are softer; their sheets are not elastic; muscovite has a different color, as does lepidolite.

Monoclinic crystal form

Lepidolite

KLi$_2$Al[(F,OH)$_2$/Si$_4$O$_{10}$]

Occurrences Mostly embedded in pegmatites and pneumatolytic veins, sometimes as raised crystals, mostly in pegmatites.

> **Hardness** 2.5–3
> **Density** 2.8–3.2
> **Luster** Pearly
> **Cleavage** Extremely perfect basal
> **Fracture** Flaky
> **Tenacity** Brittle, elastic sheets

Monoclinic crystal form

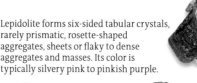

Lepidolite forms six-sided tabular crystals, rarely prismatic, rosette-shaped aggregates, sheets or flaky to dense aggregates and masses. Its color is typically silvery pink to pinkish purple.

cleavage plane
feldspar
prismatic crystal
quartz

reniform surface
scaly aggregate

Similar minerals
The color of lepidolite is extraordinarily typical; manganese-containing muscovite (alurgite) can also be pink to reddish but only occurs in metamorphic rocks.

116

Zinnwaldite

K(Li,Al,Fe)$_3$(Al,Si)$_4$O$_{10}$(OH,F)$_2$

Occurrences In pegmatites and tin and tungsten deposits.

> **Hardness** 2.5–4
> **Density** 2.9–3.3
> **Luster** Vitreous
> **Cleavage** Perfect
> **Fracture** Flaky
> **Tenacity** Flexible

Monoclinic crystal form

Zinnwaldite forms silver-gray to greenish tabular crystals with a six-sided cross-section. It also forms platy aggregates, usually accompanied by tin and tungsten ores.

crystal set

tabular crystal druse

Similar minerals
Muscovite cannot be distinguished from zinnwaldite by simple means, but paragenesis with tin minerals gives hints; biotite and phlogopite are more brown or black; lepidolite more pink.

Amber

~ $C_{10}H_{16}O+(H_2S)$

Amber is fossilized resin; it does not form crystals. It can be found in yellow, white, light to dark brown, red and even blue coarse, rounded chunks. It forms drop-shaped or spherical aggregates and flat pieces. Amber is usually covered with an opaque weathering rind; inside is often translucent to opaque. Inclusions of plant parts and living organisms (especially insects) are often present. Fresh fractures fluoresce intensively when exposed to UV light. Amber is used in many different ways in jewelry.

Occurrences *In sands, gravel, alluvial deposits.*

> ***Hardness*** 2.5
> ***Density*** 1.05–1.09
> ***Luster*** Vitreous to greasy
> ***Cleavage*** None
> ***Fracture*** Conchoidal
> ***Tenacity*** Brittle

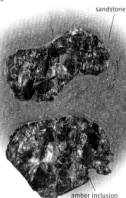

sandstone

amber inclusion

insect trapped in amber

fracture surface

weathered surface crust

platy amber with reniform surface

Ulexite

$NaCaB_5O_6(OH)_6 \cdot 5H_2O$

Occurrences In salt deposits in borax lakes, often in large beds.

> **Hardness** 2.5
> **Density** 1.95
> **Luster** Silky
> **Cleavage** Perfect
> **Fracture** Fibrous
> **Tenacity** Brittle

Ulexite is white, rarely gray, and sometimes forms acicular crystals; rather usually it forms parallel-fibrous aggregates. Often it is also coarse or forms cotton-like balls. When a piece of ulexite that has been cut perpendicular to the fiber direction is placed on top of an image, the image is projected to the surface, which is why ulexite is also known as television stone.

Triclinic crystal form

parallel-fibered

fragment

radial aggregate

fibrous

118

Wulfenite, Yellow Lead Ore
PbMoO₄

Wulfenite forms mostly raised crystals, which can be yellow, orange, red and sometimes blue. They vary from acicular, pointed bipyramidal to thick and thin tabular. Tabular crystals often have a sandwich-like structure.

Occurrences In the oxidation zone of lead deposits.

> **Hardness** 3
> **Density** 6.7–6.9
> **Luster** Adamantine to resinous
> **Cleavage** Poor pyramidal cleavage
> **Fracture** Conchoidal
> **Tenacity** Brittle

Similar minerals
The appearance (especially crystal form and orange color) and occurrence of wulfenite with other lead and zinc oxidation minerals do not permit any confusion; calcite effervesces when dabbed with diluted hydrochloric acid; vanadinite always shows six-sided forms; the same applies to mimetesite.

Tetragonal crystal form

bipyramidal crystals

limonite

prismatic crystals

calcite

119

resinous

Vanadinite
Pb₅[Cl/(VO₄)₃]

$Pb_5[Cl/(VO_4)_3]$

Vanadinite mostly forms raised acicular, prismatic to tabular crystals, as well as radial and spherical aggregates, and is rarely coarse. Red is the most common color of vanadinite, but it also forms gray or yellow to brown crystals.

Similar minerals
Apatite is harder; pyromorphite and mimetesite are not red; brown or yellow vanadinite cannot be distinguished from mimetesite or pyromorphite by simple means; red wulfenite does not form hexagonal crystals.

Hexagonal crystal form

six-sided end face

descloizite

six-sided column

prismatic crystal

reniform romanechite

120

Phosgenite, Horn Lead
$Pb_2[Cl_2/CO_3]$

Phosgenite forms acicular, prismatic to tabular crystals, which can be colorless or yellowish to brownish in color. The crystals are sometimes rich in faces.

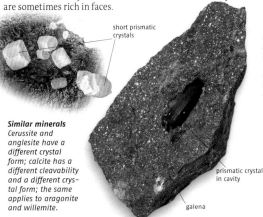

short prismatic crystals

Similar minerals
Cerussite and anglesite have a different crystal form; calcite has a different cleavability and a different crystal form; the same applies to aragonite and willemite.

prismatic crystal in cavity

galena

Occurrences *In the oxidation zone of lead deposits, in ancient lead slags.*

> **Hardness** 3
> **Density** 6–6.3
> **Luster** *Greasy to adamantine*
> **Cleavage** *Perfect*
> **Fracture** *Conchoidal*
> **Tenacity** *Brittle*

Tetragonal crystal form

Anglesite
$PbSO_4$

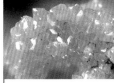

Anglesite forms colorless to white or, more rarely, yellow or greenish raised crystals. These are tabular, prismatic, bipyramidal or acicular. Anglesite is also granular, crusty or coarse.

greenish anglesite

corroded galena

prismatic crystal

Similar minerals
Barite has a much better cleavage; cerussite, in contrast to anglesite, often shows knee-shaped twins and star-shaped triplets.

Occurrences *In the oxidation zone of lead deposits, often as the first formation during the weathering of galena.*

> **Hardness** 3
> **Density** 6.3
> **Luster** *Vitreous to greasy*
> **Cleavage** *Visible basal cleavage*
> **Fracture** *Conchoidal*
> **Tenacity** *Brittle*

Orthorhombic crystal form

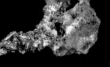

Silver (Native)
Ag

Occurrences *In hydrothermal veins; rarely primary, mostly secondary through cementation formation, on fissured surfaces in black shales.*

> **Hardness** 2.5–3
> **Density** 9.6–12
> **Luster** Metallic, sometimes dull from tarnish
> **Cleavage** None
> **Fracture** Hackly
> **Tenacity** Brittle, very ductile, can be hammered into small plates

Native silver is a fresh silver-white color, but it often is yellowish or blackish. It rarely forms well-formed cubic crystals, rather, more frequently it forms wires, coils, sheets, dendritic and skeletal aggregates. Native silver used to be a sought-after silver ore. Today it is far too rare for industrial mining.

Isometric crystal form

122

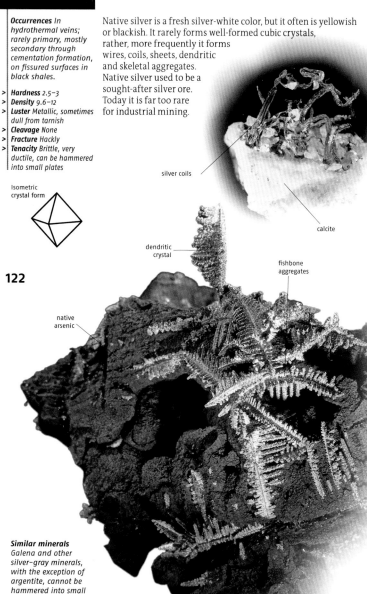

silver coils

calcite

dendritic crystal

fishbone aggregates

native arsenic

Similar minerals *Galena and other silver-gray minerals, with the exception of argentite, cannot be hammered into small plates; argentite has a dark streak.*

Duftite

CuPbAsO₄OH

$CuPbAsO_4OH$

Duftite forms yellowish-green thick tabular crystals, crystal druse, and, more often, spherical aggregates or crusty coatings. It occurs where lead and copper ores weather at the same time.

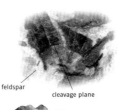

limestone

crystal druse

limonite

spherical crystal aggregates

Similar minerals
Olivenite has a different crystal form; conichalcite is more apple green; cornwallite is more emerald green. Chrysocolla is more blue-green; pyromorphite and mimetesite have a different crystal form.

Occurrences *In the oxidation zone of lead-copper deposits.*

> **Hardness** 3
> **Density** 6.4
> **Luster** Vitreous, greasy
> **Cleavage** None recognizable
> **Fracture** Uneven
> **Tenacity** Brittle

Orthorhombic crystal form

Astrophyllite

$(K,Na)_3(Fe,Mn)_7Ti_2[(O,OH)_7/Si_8O_{24}]$

Astrophyllite forms yellowish to greenish olive tabular crystals or platy aggregates. Frequently, it grows together in a radial pattern to form rosettes. Raised crystals are very rare. In most cases astrophyllite is embedded in the rock.

feldspar

cleavage plane

radial aggregate

feldspar

Similar minerals
Mica minerals are not brittle, and their luster isn't as metallic; aegirine has a different cleavability with a cleavage angle of about 90°.

Occurrences *In alkali rocks and their pegmatites.*

> **Hardness** 3
> **Density** 3.3–3.4
> **Luster** Metallic vitreous
> **Cleavage** Perfect
> **Fracture** Flaky
> **Tenacity** Brittle

Triclinic crystal form

Calcite
CaCO₃

Occurrences Rock-
forming component
(limestone, marble), in
fissures, in hydrother-
mal veins.

> **Hardness** 3
> **Density** 2.6–2.8
> **Luster** Vitreous
> **Cleavage** Very perfect
>   basal rhombohedral
> **Fracture** Splintery to
>   conchoidal
> **Tenacity** Brittle

Calcite has more crystal forms than any other mineral. It forms
scalenohedron, rhombohedron or prisms with bases and
numerous combinations. Its habit can be prismatic, isometric,
lenticular, acicular, or thick and thin tabular. The crystals are
mostly translucent to transparent white, but they can also be
yellow, brown, green, purple, red or black.

Trigonal crystal system

Similar minerals
Calcite effervesces
when dabbed with
diluted cold hydro-
chloric acid, dolomite
only with hot hydro-
chloric acid; quartz
is harder; gypsum
is softer; anhydrite
has a cleavability
with right cleavage
angles and does
not effervesce with
hydrochloric acid.

thin tabular crystals

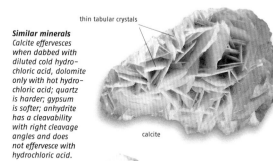

calcite

124

prismatic crystal

six-sided end face

Calcite often forms twins (natural intergrowth of two crystals), which can be heart or butterfly shaped. Its defining twin characteristic is their different angles, which cannot occur on single crystals, and depending on whether the individual crystals are scalenohedra, rhombohedra or prismatic, completely different shapes can be created.

hematite coating

rhombohedron

Occurrences *Rock-forming component (limestone, marble), in fissures, in hydrothermal veins.*

> ***Hardness*** *3*
> ***Density*** *2.6–2.8*
> ***Luster*** *Vitreous*
> ***Cleavage*** *Very perfect basal rhombohedral*
> ***Fracture*** *Splintery to conchoidal*
> ***Tenacity*** *Brittle*

Trigonal crystal system

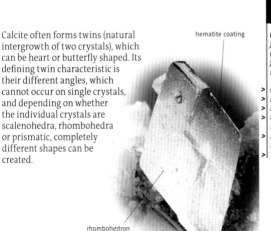

crystal druse

heart twinning

re-entrant angle

Calcite
$CaCO_3$

Occurrences Rock-forming component (limestone, marble), in fissures, in hydrothermal veins.

> **Hardness** 3
> **Density** 2.6–2.8
> **Luster** Vitreous
> **Cleavage** Very perfect basal rhombohedral
> **Fracture** Splintery to conchoidal
> **Tenacity** Brittle

Trigonal crystal system

Calcite often also forms radial, spherical and reniform aggregates in the form of stalactites, as vein fillings or a rock-forming component in limestone and coarse marble. Clear fragments of calcite allow the observation of a special property: birefringence. When you lay such pieces on lined paper, all lines appear twice. Such fragments are also called double refraction calcspar or, because of where it was discovered, Iceland spar.

fragment

birefringence

double-refraction calcspar

126

calcite sphere

basalt

The occurrence of calcite is varied: Crystals in druses of ore veins, bubble cavities of volcanic rocks, in fissures and in druses of carbonate rocks, as a gangue in many hydrothermal veins; magmatic rock-forming component in carbonatites, sedimentary rock-forming component in limestones, in limestone tuffs, in marls, as a binding agent in sandstones, metamorphic rock-forming component in marbles.

Occurrences *Rock-forming component (limestone, marble), in fissures, in hydrothermal veins.*

> ***Hardness*** *3*
> ***Density*** *2.6–2.8*
> ***Luster*** *Vitreous*
> ***Cleavage*** *Very perfect basal rhombohedral*
> ***Fracture*** *Splintery to conchoidal*
> ***Tenacity*** *Brittle*

Trigonal crystal system

rounded crystals

delessite (chlorite)

scalenohedral

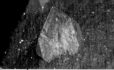

Barite
BaSO$_4$

Barite is mostly white or translucent to transparent, but it can also be pink, yellowish, brownish or even blue. Its crystals are tabular, rarely prismatic. Fan-shaped and tap comb-like to spherical aggregates are common. In sands, you can also find blossom-shaped aggregates (barite roses), coarse barite is splintery; very typical is its noticeably high specific weight.

barite rose

Orthorhombic crystal form

tabular crystals

cerussite

128

barite-colored by iron

Celestine, Celestite

SrSO$_4$

Celestine is mostly white or translucent to transparent, but it can also be yellow, brownish or even blue. Its crystals are tabular with long or short prisms. Coarse celestine is splintery, and aggregates can be platy to radial.
Sometimes parallel-fibrous crevasse fillings occur in limestones and marls.

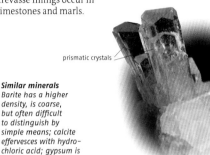

prismatic crystals

Similar minerals
Barite has a higher density, is coarse, but often difficult to distinguish by simple means; calcite effervesces with hydrochloric acid; gypsum is much softer.

Occurrences *In hydrothermal veins, as a filling in crevasses and druse in limestones and marls.*

> ***Hardness*** 3–3.5
> ***Density*** 3.9–4
> ***Luster*** Vitreous, pearly on cleavage planes
> ***Cleavage*** Perfect basal, two additional cleavage directions are much poorer
> ***Fracture*** Uneven
> ***Tenacity*** Brittle

Orthorhombic crystal form

crystal clusters

prismatic crystals

calcite

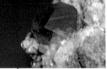

Cerussite, White Lead Ore
$PbCO_3$

Occurrences In the oxidation zone of lead deposits.

> **Hardness** 3–3.5
> **Density** 6.4–6.6
> **Luster** Greasy to adamantine
> **Cleavage** Poorly recognizable
> **Fracture** Conchoidal
> **Tenacity** Brittle

The mostly white to colorless crystals of cerussite are prismatic, isometric, and tabular. Often knee-shaped twins appear. The multiple formation of twins results in star-shaped and lattice-shaped structures. Additionally, cerussite forms reniform aggregates and can be crusty, earthy, coarse. Crystals colored black by residual galena are also called black lead ore.

tabular crystals

Orthorhombic crystal form

prismatic crystals

Similar minerals
In contrast to cerussite, calcite and aragonite effervesce with diluted hydrochloric acid; the characteristic twinning distinguishes cerussite from anglesite.

greasy luster

Anhydrite

CaSO₄

Anhydrite is the most abundant rock-forming component; forming dense white, yellowish or blue splintery to granular masses. Thick to thin tabular, isometric to prismatic crystals are less common. Water absorption can transform it into gypsum, often resulting in highly folded gypsum aggregates.

Occurrences *In salt deposits, sedimentary rocks, as separate rock bodies, in hydrothermal veins as gangue, in alpine-type fissures.*

> **Hardness** *3–3.5*
> **Density** *2.98*
> **Luster** *Vitreous*
> **Cleavage** *Perfect, cuboid cleavage structures with right angles*
> **Fracture** *Splintery*
> **Tenacity** *Brittle*

blue coarse anhydrite

Similar minerals
The characteristic rectangular cleavability makes anhydrite unmistakable; calcite effervesces when dabbed with diluted hydrochloric acid; gypsum is much softer and has a completely different cleavability.

Orthorhombic crystal form

cleavage

cuboid

vitreous luster

Adamite
Zn₂[OH/AsO₄]

Occurrences *In the oxidation zone of zinc deposits, which also carry arsenic-containing primary minerals.*

> ***Hardness*** 3.5
> ***Density*** 4.3–4.5
> ***Luster*** Vitreous
> ***Cleavage*** Perfect, but mostly not recognizable
> ***Fracture*** Conchoidal
> ***Tenacity*** Brittle

Adamite forms prismatic to acicular crystals, radial or sun-shaped aggregates, is often crusty, reniform, and rarely coarse. Pure adamite is colorless to yellow but can be more or less green (cuprian adamite) when it contains copper. Cobalt and manganese contents produces a purple color.

Orthorhombic crystal form

cuprian adamite

limonite

132

sun-shaped aggregate

prismatic crystals

Similar minerals
Olivenite is always dark green; anglesite and cerussite have a different crystal form.

Paradamite

$Zn_2[OH/AsO_4]$

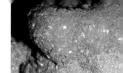

aggregate of thick tabular crystals

Paradamite forms yellow tabular crystals, often rounded. It mostly forms raised crystals in druses in limonite, rarely is it coarse.

lenticular crystals

Occurrences *In the oxidation zone of zinc deposits, which also carry arsenic-containing primary minerals.*

> ***Hardness*** *3.5*
> ***Density*** *4.55*
> ***Luster*** *Vitreous*
> ***Cleavage*** *Perfect*
> ***Fracture*** *Uneven*
> ***Tenacity*** *Brittle*

Similar minerals
Adamite has a different crystal form; yellowish calcite effervesces in contrast to paradamite when dabbed with diluted hydrochloric acid; barite has a different crystal form.

Triclinic crystal form

vitreous luster

Ludlamite

$Fe_3[PO_4]_2 \cdot 4\,H_2O$

regular association

Ludlamite forms light green to green octahedron-like, thick to thin tabular crystals, as well as rosette-like aggregates and coarse, splintery, easily cleavable masses. It often occurs together with vivianite.

fairfieldite

Occurrences *In phosphate pegmatites, on hydrothermal ore deposits.*

> ***Hardness*** *3–4*
> ***Density*** *3.1*
> ***Luster*** *Vitreous*
> ***Cleavage*** *Perfect basal*
> ***Fracture*** *Uneven*
> ***Tenacity*** *Brittle*

thick sheets

Monoclinic crystal form

pyrite

Aragonite
CaCO₃

Occurrences *In the oxidation zone, in druses and in fissures of effusive rocks, embedded in clays (mostly triplets here), in the deposits of hot springs.*

> **Hardness** 3.5–4
> **Density** 2.95
> **Luster** *Vitreous*
> **Cleavage** *Indiscernible*
> **Fracture** *Conchoidal*
> **Tenacity** *Brittle*

The white, colorless, rarely gray to red crystals of aragonite are mostly acicular, prismatic, spatulate. Triplets resemble hexagonal prisms, which are particularly common in Spain. There are also radial, granular aggregates. Vermicular, coral-like white formations are called "flowers of iron." They are primarily found in the Styrian Erzberg in Austria.

limonite

acicular

Orthorhombic crystal form

134

Similar minerals
Calcite differs from aragonite by its cleavability; all other minerals are distinguishable by the hydrochloric acid test, as in contrast to aragonite they do not effervesce when dabbed with diluted hydrochloric acid.

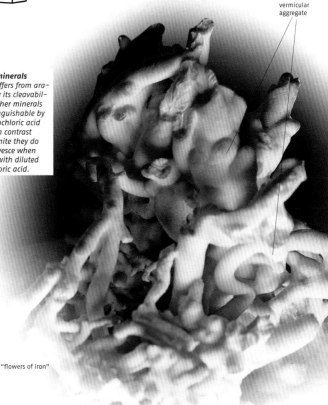

vermicular aggregate

"flowers of iron"

Gyrolite

$NaCa_{16}(Si_{23}Al)O_{60}(OH)_8 \cdot 14\, H_2O$

Gyrolite forms spherical aggregates of platy to tabular crystals. Its color varies from white to yellow to green, brown and even black.

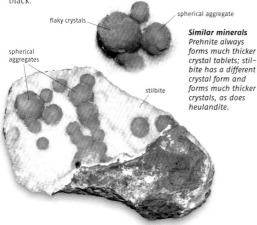

flaky crystals

spherical aggregate

spherical aggregates

stilbite

Occurrences In cavities within volcanic rocks.

> **Hardness** 3–4
> **Density** 2.34–2.45
> **Luster** Vitreous
> **Cleavage** Perfect
> **Fracture** Uneven
> **Tenacity** Brittle

Similar minerals Prehnite always forms much thicker crystal tablets; stilbite has a different crystal form and forms much thicker crystals, as does heulandite.

Hexagonal crystal form

Strunzite

$MnFe_2[OH/PO_4]_2$

limonite

Strunzite forms straw-yellow acicular to capillary crystals, which very rarely can be prismatic. Fibrous to radial aggregates are often found in cavities of transformed phosphate minerals.

acicular crystals

acicular crystals

Occurrences In phosphate-bearing pegmatites, in phosphate-rich brown iron deposits.

> **Hardness** 4
> **Density** 2.52
> **Luster** Vitreous
> **Cleavage** None
> **Fracture** Uneven
> **Tenacity** Brittle

limonite

Similar minerals Cacoxenite is more golden yellow, but is often indistinguishable from strunzite by simple means; beraunite is more orange and not so straw yellow.

Triclinic crystal form

Variscite
Al[PO₄] · 2H₂O

Occurrences *In fissures of rocks rich in aluminum.*

> **Hardness** 4–5
> **Density** 2.52
> **Luster** Vitreous to waxy
> **Cleavage** None
> **Fracture** Conchoidal
> **Tenacity** Brittle

Variscite seldom forms white to colorless crystals, rather they are mostly radial, spherical aggregates, with crusty coatings and whitish to green coarse, dense masses. The latter type are used for jewelry purposes.

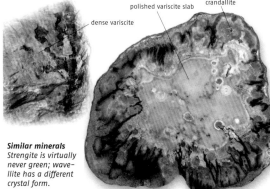

polished variscite slab
crandallite
dense variscite

Orthorhombic crystal form

Similar minerals
Strengite is virtually never green; wavellite has a different crystal form.

136

Strengite
Fe[PO₄] · 2 H₂O

Occurrences *In phosphorus-containing brown iron deposits and phosphate pegmatites, where it is formed through the weathering of other phosphate minerals.*

> **Hardness** 3–4
> **Density** 2.87
> **Luster** Vitreous
> **Cleavage** Perfect basal
> **Fracture** Conchoidal
> **Tenacity** Brittle

Strengite forms blue-purple to pink-purple tabular to isometric crystals that are often rich in faces, as well as radial and spherical aggregates, crusts, and coatings. It occurs together with other secondary phosphate minerals such as cacoxenite or phosphosiderite.

cacoxenite
spherical aggregate
radial-fibered aggregates
rockbridgeite

Orthorhombic crystal form

Similar minerals
Phosphosiderite has a different crystal form, but is not easily distinguishable from strengite in radial aggregates; amethyst, which is very similar in color, is much harder.

Pyromorphite, Green Lead Ore

$Pb_5[Cl/(PO_4)_3]$

Pyromorphite forms prismatic to acicular crystals, often barrel-shaped due to curved prismatic surfaces ("Emser Barrels"). In addition, it forms radial, reniform aggregates. It is also crusty, earthy, rarely coarse. Its color varies from green to white or yellow to brown (brown lead ore).

Occurrences *In the oxidation zone of lead deposits, especially in the upper parts exposed to weathering.*

> ***Hardness*** 3.5–4
> ***Density*** 6.7–7
> ***Luster*** *Greasy*
> ***Cleavage*** *None*
> ***Fracture*** *Conchoidal*
> ***Tenacity*** *Brittle*

Similar minerals
Mimetite is often difficult to distinguish from pyromorphite by simple means, but minerals containing arsenic as accompanying minerals can give an indication of the presence of mimetite.

brown lead ore

calcite

Hexagonal crystal form

137

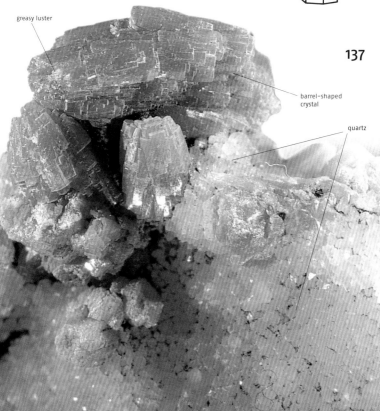

greasy luster

barrel-shaped crystal

quartz

Mimetite
Pb$_5$[Cl/(AsO$_4$)$_3$]

Occurrences In the oxidation zone of lead deposits, which also carry arsenic-containing minerals.

> **Hardness** 3.5–4
> **Density** 7.1
> **Luster** Adamantine to greasy
> **Cleavage** None
> **Fracture** Conchoidal
> **Tenacity** Brittle

Hexagonal crystal form

Mimetite forms prismatic to acicular crystals, often barrel-shaped to spherical in shape due to curved prismatic surfaces. Also as radial, reniform aggregates and crusty, earthy, rarely coarse, masses. Its color varies from colorless, white, yellow orange to brown and gray.

Similar minerals
Apatite is harder; vanadinite and pyromorphite cannot be distinguished by simple means, but the paragenesis of mimetite with arsenic-containing minerals gives clues; vanadinite is mostly red, this color practically does not occur with mimetite.

spherically curved crystal

limonite

138

limonite

sheaf-like crystals

Stilbite

$Ca[Al_2Si_7O_{18}] \cdot 7\,H_2O$

Stilbite always forms twinned prismatic to tabular crystals, often intergrown into clusters of
sheaves but also spherical, radial aggregates, it is almost always raised. The color varies from colorless, white to yellow to brown.

Similar minerals
The typical crystal shape of stilbite leaves hardly any room for confusion.

calcite

doubly
terminated crystal

Occurrences *In druses and in fissures in igneous rocks, in alpine-type fissures, in ore veins.*

> **Hardness** 3.5–4
> **Density** 2.1–2.2
> **Luster** Vitreous, pearly on cleavage planes
> **Cleavage** Perfect
> **Fracture** Uneven
> **Tenacity** Brittle

Monoclinic crystal form

sheaf-like crystals

Rhodochrosite, Raspberry Spar

$MnCO_3$

Occurrences *In hydrothermal veins, in the oxidation zone of iron-manganese deposits, as lenses and deposits in metamorphic rocks.*

> **Hardness** *3.5–4*
> **Density** *3.3–3.6*
> **Luster** *Vitreous*
> **Cleavage** *Perfect, rhombohedral*
> **Fracture** *Uneven*
> **Tenacity** *Brittle*

Rhodochrosite is pink to red; in rare cases, it is yellowish. It forms rhombohedral or scalenohedral habits, which are often rounded ("rice grain-shaped"). Often you can also find spherical, reniform and radial aggregates, it is also stalactitic, crusty, coarse. Beautifully colored, dense rhodochrosite is used to make jewelry.

intergrowth of curved rhombohedron

Trigonal crystal system

scalenohedral

rounded surfaces

140

limonite

Similar minerals
In contrast to rhodochrosite, calcite reacts with diluted cold hydrochloric acid; at times, rhodochrosite cannot be distinguished by simple means from manganese-containing dolomite, which can also be pink.

Dolomite

CaMg(CO₃)₂

The white, rarely colorless, yellowish, brownish or rarely pink-colored dolomite crystals are usually only characterized by the basal rhombohedron, often typically curved in the shape of a saddle. Dolomite is often coarse, commonly a component in rock-forming. It effervesces only when dabbed with hot concentrated hydrochloric acid (caution!).

Occurrences *In hydrothermal veins as gangue and in druses, a component in rock-forming, crystals often present in fissures in dolomite rocks.*

> **Hardness** *3.5–4*
> **Density** *2.85–2.95*
> **Luster** *Vitreous*
> **Cleavage** *Perfect rhombohedral*
> **Fracture** *Splintery*
> **Tenacity** *Brittle*

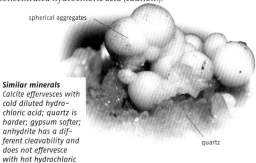

spherical aggregates

Similar minerals
Calcite effervesces with cold diluted hydrochloric acid; quartz is harder; gypsum softer; anhydrite has a different cleavability and does not effervesce with hot hydrochloric acid.

quartz

Trigonal crystal system

chalcopyrite

saddle-shaped curved rhombohedra

quartz crystals

Phosphosiderite

$Fe[PO_4] \cdot 2H_2O$

Occurrences *In phosphate-bearing pegmatites as a conversion product of primary phosphate minerals.*

> ***Hardness*** 3.5–4
> ***Density*** 2.76
> ***Luster*** Vitreous
> ***Cleavage*** Perfect
> ***Fracture*** Uneven
> ***Tenacity*** Brittle

Monoclinic crystal form

Phosphosiderite forms colorless, white, pink to purple, thick to thin tabular crystals. Often twins are formed with intruding angles. Even more frequent are radial aggregates and crusty and warty coatings.

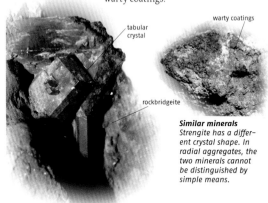

tabular crystal

warty coatings

rockbridgeite

Similar minerals
Strengite has a different crystal shape. In radial aggregates, the two minerals cannot be distinguished by simple means.

142

Scorodite

$Fe[AsO_4] \cdot 2 H_2O$

Occurrences *In the oxidation zone of arsenic-rich deposits, often as a direct conversion product of arsenopyrite.*

> ***Hardness*** 3.5–4
> ***Density*** 3.1–3.3
> ***Luster*** Greasy vitreous luster
> ***Cleavage*** Hardly visible
> ***Fracture*** Conchoidal
> ***Tenacity*** Brittle

Orthorhombic crystal form

Scorodite forms colorless, white, greenish, blue tabular to bipyramidal crystals, as well as radial and spherical aggregates, crusts, or coatings together with other secondary arsenic minerals (e.g. carminite).

quartz

prismatic crystals

carminite

bipyramidal crystals

Similar minerals
If crystal form and paragenesis with other arsenic-containing minerals are observed, there is no possibility of confusion.

Siderite, Iron Spar
FeCO₃

The yellowish, yellowish brown to dark brown, sometimes bluish to reddish tarnished siderite crystals usually only show rhombohedron forms. On the other hand, dense six-sided crystals or scalenohedra are much rarer, as are reniform, spherical aggregates. Siderite is usually intergrown with other ore minerals and is coarse and splintery.

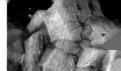

Occurrences *In pegmatites and volcanic rocks, in hydrothermal veins, in metasomatically modified limestones, in sediments.*

> **Hardness** 4–4.5
> **Density** 3.7–3.9
> **Luster** Vitreous
> **Cleavage** Perfect rhombohedral
> **Fracture** Splintery
> **Tenacity** Brittle

spherical aggregate
(spheroidal siderite)

Similar minerals
In contrast to siderite, calcite effervesces with diluted hydrochloric acid; sphalerite has a different cleavability.

Trigonal crystal system

143

rock cyrstals

cleavage

rhombohedral crystals

Magnesite
MgCO₃

$MgCO_3$

Occurrences *Large displacement bodies in dolomites, in talc schists, in fissures and in veins in serpentine.*

> **Hardness** 4–4.5
> **Density** 3
> **Luster** *Vitreous*
> **Cleavage** *Very perfect rhombohedral*
> **Fracture** *Splintery*
> **Tenacity** *Brittle*

Magnesite forms mostly hexagonal plates and, rarely, rhombohedral crystals. It often forms white, yellowish to brownish coarse, granular and splintery masses, sometimes mottled black and white. This so-called pinolite magnesite is also used for arts and crafts.

Trigonal crystal system

Similar minerals
In contrast to magnesite, calcite effervesces with diluted cold hydrochloric acid; dolomite is somewhat softer, but often cannot be distinguished from magnesite by simple means. Pinolite magnesite is unmistakable.

tabular crystals

144

pinolite magnesite with typical black and white appearance

white magnesite

colored black by carbon

Wavellite

$Al_3[(OH)_3/(PO_4)_2] \cdot 5\ H_2O$

Wavellite forms acicular to prismatic crystals, much more frequent are radial aggregates and crusty and warty coatings, sun-shaped aggregates often appear on fissured surfaces. The color varies from colorless to white, yellow to green.

Occurrences *In fissures of siliceous slate, decomposed granite, limestone.*

> **Hardness** 4
> **Density** 2.3–2.4
> **Luster** Vitreous
> **Cleavage** Not visible due to acicular habit
> **Fracture** Uneven
> **Tenacity** Brittle

Monoclinic crystal form

spherical aggregates

radial aggregates

siliceous shale

Similar minerals *Natrolite and prehnite are harder; calcite and aragonite are more resistant to dabbing with diluted hydrochloric acid than wavellite.*

Heulandite

$Ca[Al_2Si_7O_{18}] \cdot 6\ H_2O$

Heulandite forms thin to thick tabular crystals, radial to spherical aggregates, always raised. Heulandite can come in many colors, from colorless to white, yellow to red and chocolate brown.

Occurrences *In druses in pegmatites, in ore veins, in bubble cavities in volcanic rocks.*

> **Hardness** 3.5–4
> **Density** 2.2
> **Luster** Vitreous, pearly on cleavage planes
> **Cleavage** perfect, one cleavage plane
> **Fracture** Uneven
> **Tenacity** Brittle

Monoclinic crystal form

thick tabular crystal

chlorite

colored red by iron

Similar minerals *Stilbite, phillipsite and chabasite have a different crystal form; in contrast to heulan-dite, calcite effervesces when dabbed with diluted hydrochloric acid; apophyllite has a different crystal form.*

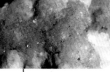

Phillipsite
KCa[Al$_3$Si$_5$O$_{16}$] · 6 H$_2$O

prismatic crystals

Occurrences *Raised in bubble cavities of volcanic rocks.*

> **Hardness** *4–4.5*
> **Density** *2.2*
> **Luster** *Vitreous*
> **Cleavage** *Discernible*
> **Fracture** *Uneven*
> **Tenacity** *Brittle*

Phillipsite always forms twins, mostly prismatic twins and quadruplets, but also duodecuplets, which look like rhombic dodecahedrons. It is almost always raised and colorless to white, rarely yellowish or reddish.

Monoclinic crystal form

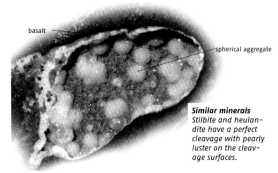

basalt

spherical aggregate

Similar minerals
Stilbite and heulandite have a perfect cleavage with pearly luster on the cleavage surfaces.

Harmotome
Ba[Al$_2$Si$_6$O$_{16}$] · 6 H$_2$O

Occurrences *In the cavities of volcanic rocks, in hydrothermal ore deposits.*

> **Hardness** *4.5*
> **Density** *2.44–2.5*
> **Luster** *Vitreous*
> **Fracture** *Conchoidal*
> **Tenacity** *Brittle*

Harmotome almost always forms cruciform twins and prismatic crystals. It is always raised and colorless, white or pink.

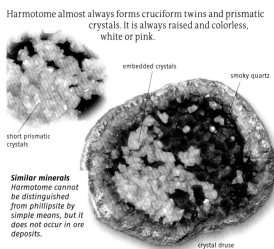

embedded crystals

smoky quartz

short prismatic crystals

crystal druse

Monoclinic crystal form

Similar minerals
Harmotome cannot be distinguished from phillipsite by simple means, but it does not occur in ore deposits.

Kyanite, Disthene, Cyanite
$Al_2[O/SiO_4]$

Kyanite forms white, gray to black but especially blue, columnar crystals and aggregates, which are almost always embedded. A special characteristic is its large difference in hardness: kyanite has a hardness of 4 in the longitudinal direction, meaning it can easily be scored with a knife. However, this is not possible in the transverse direction because here it has a hardness of 7.

> ***Occurrences*** *Embedded into metamorphic rocks, gneisses, mica schists.*
> ***Hardness*** *4 (vertical)–7 (horizontal)*
> ***Density*** *3.6–3.7*
> ***Luster*** *Vitreous*
> ***Cleavage*** *Perfect*
> ***Fracture*** *Uneven*
> ***Tenacity*** *Brittle*

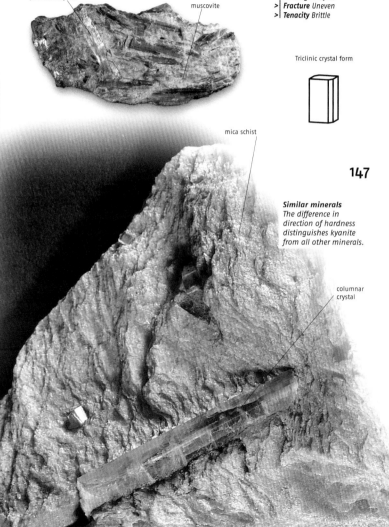

prismatic crystal

muscovite

mica schist

Triclinic crystal form

147

Similar minerals
The difference in direction of hardness distinguishes kyanite from all other minerals.

columnar crystal

Fluorite, Fluorspar
CaF₂

CaF_2

Occurrences In pegmatites, hydrothermal veins, alpine-type fissures.

> **Hardness** 4
> **Density** 3.1–3.2
> **Luster** Vitreous
> **Cleavage** Perfect octahedral
> **Fracture** Uneven
> **Tenacity** Brittle

Fluorite often forms beautifully formed crystals, most often cubes, more rarely octahedrons, very rarely rhombic dodecahedrons and other forms. Its color varies from colorless to white, yellow, honey brown, green, blue, pink, red and black purple. Several colors can even appear in one piece. Rarer are radial, reniform aggregates or spherical formations.

purple fluorite octahedron

feldspar

Isometric crystal form

Similar minerals
Fluorite differs from apatite in crystal form and cleavability and from calcite and quartz in hardness; halite is water-soluble and tastes salty.

148

rhombic dodecahedron surface

cube surface

white fluorite

Fluorite is a very widespread mineral. In hydrothermal veins, it occurs as a kind of gangue, where some very large crystals can be found in cavities. Crystals can also be found in druses and in fissures in limestones, in fissures of silicate rocks, in alpine-type fissures. Coarse layers of fluorite are found in sedimentary rocks. Multicolored banded fluorite is also used to make craft items.

green octahedron

Occurrences In pegmatites, hydrothermal veins, alpine-type fissures.

> *Hardness* 4
> *Density* 3.1–3.2
> *Luster* Vitreous
> *Cleavage* Perfect octahedral
> *Fracture* Uneven
> *Tenacity* Brittle

Isometric crystal form

pink octahedrons from an alpine-type fissure

149

rock crystal

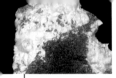

Tobermorite

$Ca_5Si_6O_{16}(OH)_2 \cdot 4\,H_2O$

Occurrences In calcareous xenoliths, in druses in basalts, in metamorphic manganese deposits.

> **Hardness** 4.5
> **Density** 2.4
> **Luster** Vitreous to silky
> **Cleavage** Poorly recognizable
> **Fracture** Fibrous
> **Tenacity** Brittle

Tobermorite forms colorless to white acicular crystals, crystal clusters, and fibrous and radial aggregates. Artificially produced, it is an important component of pervious concrete.

acicular crystals

Monoclinic crystal form

silky luster

radial fibrous aggregate

Similar minerals
Natrolite usually shows thicker crystals, often with end faces; aragonite effervesces when dabbed with diluted hydrochloric acid; tremolite occurs only in metamorphic rocks.

Smithsonite, Zinc Spar
$ZnCO_3$

Smithsonite is multicolored.
It can be colorless, white,
yellow, green, brown, pink
or blue. It forms
scalenohedra and
rhombohedron, often
rounded, rice
grain-shapes. Its
aggregates are
reniform, stalactitic,
scaly, coarse.

rhombohedron-shaped crystal

Occurrences *In the
oxidation zone of zinc
deposits.*

> ***Hardness*** 5
> ***Density*** 4.3–4.5
> ***Luster*** Vitreous
> ***Cleavage*** Perfect, rhom-
> bohedral
> ***Fracture*** Uneven
> ***Tenacity*** Brittle

Trigonal crystal system

151

Similar minerals
*In contrast to zinc
spar, calcite effervesces
with diluted hydro-
chloric acid; dolomite
does not normally
occur in the oxidation
zone of zinc deposits.*

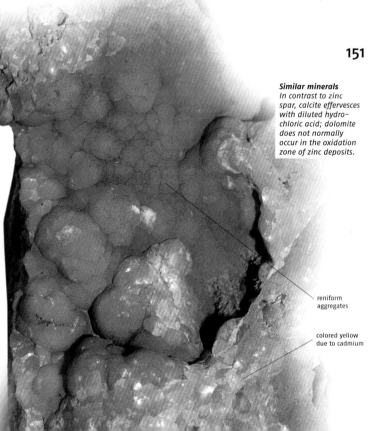

reniform
aggregates

colored yellow
due to cadmium

Colemanite
$Ca[B_3O_4(OH)_3] \cdot H_2O$

Occurrences In borax lakes and corresponding sediments.

> **Hardness** 4.5
> **Density** 2.4
> **Luster** Vitreous
> **Cleavage** Perfect
> **Fracture** Uneven
> **Tenacity** Brittle

Monoclinic crystal form

Colemanite forms white prismatic to tabular crystals. It can be granular, columnar and coarse.

tabular crystal

colored by limonite

Similar minerals Borax and soda ash are softer; calcite and aragonite effervesce when dabbed with diluted hydrochloric acid; barite and celestine are much heavier; anhydrite shows a rectangular cleavability.

Wollastonite
$Ca_3[Si_3O_9]$

Occurrences In metamorphic limestones, in skarn deposits.

> **Hardness** 4.5
> **Density** 2.44–2.5
> **Luster** Vitreous
> **Fracture** Conchoidal
> **Tenacity** Brittle

Triclinic crystal form

Wollastonite rarely forms tabular crystals, mostly fibrous, radial aggregates. It is often coarsely fibrous. It also forms dendritic growths on fissured surfaces.

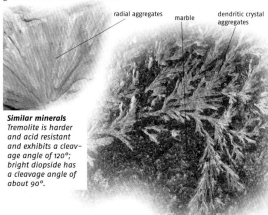

radial aggregates

marble

dendritic crystal aggregates

Similar minerals Tremolite is harder and acid resistant and exhibits a cleavage angle of 120°; bright diopside has a cleavage angle of about 90°.

Chabazite

$Ca[Al_2Si_4O_{12}] \cdot 12H_2O$

Chabazite forms colorless, white, yellow cube-like rhombohedra, often twins, always raised. It is sometimes found as flat six-sided pyramids; this variety is known as phacolite.

Occurrences In bubble cavities of volcanic rocks and cavities of pegmatites, in druses and fissures of ore veins, in alpine-type fissures.

> *Hardness* 4.5
> *Density* 2.08
> *Luster* Vitreous
> *Cleavage* Indiscernible
> *Fracture* Uneven
> *Tenacity* Brittle

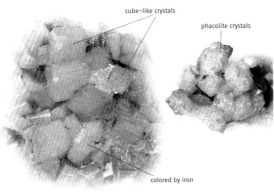

cube-like crystals

phacolite crystals

colored by iron

Trigonal crystal system

153

Similar minerals
Limestone differs from chabasite by its cleavability and effervesces when dabbed with diluted hydrochloric acid; fluorite also has a clear cleavability in contrast to chabasite.

basalt

cube-like crystals

Serpentine, Antigorite, Chrysotile

$Mg_6[(OH)_8/Si_4O_{10}]$

Occurrences Rock-forming component in serpentinites, chrysotile in the fissures of this rock.

> **Hardness** 3–4
> **Density** 2.5–2.6
> **Luster** Greasy to silky
> **Cleavage** Usually not recognizable due to fine-grained habit
> **Fracture** Conchoidal to fibrous
> **Tenacity** Brittle

Serpentine occurs in two different species: The yellow to dark green antigorite is flaky, usually very fine-grained. The yellowish green silky chrysotile is fine-grained with a capillary habit (chrysotile asbestos). Beautifully colored antigorite is used for handicraft purposes and is cut into gemstones and art objects.

Monoclinic crystal form

chrysotile asbestos

154

Similar minerals
Talc is softer; hornblende asbestos (fine-grained hornblende minerals) is brittle in contrast to chrysotile. Jadeite and nephrite are harder and not so yellowish.

magnetite

greasy luster

fine-grained antigorite

Okenite

$CaSi_2O_4(OH)_2 \cdot H_2O$

Okenite forms acicular, seldom long tabular crystals, fine acicular spheres ("cotton balls") and radial aggregates, spherical bundles and fine fibrous aggregates. It is always a colorless white. Green, blue or red okenite bundles are always fakes.

Occurrences In cavities within volcanic rocks.

> **Hardness** 4.5–5
> **Density** 2.3
> **Luster** Vitreous
> **Cleavage** None recognizable
> **Fracture** Uneven, splintery
> **Tenacity** Brittle

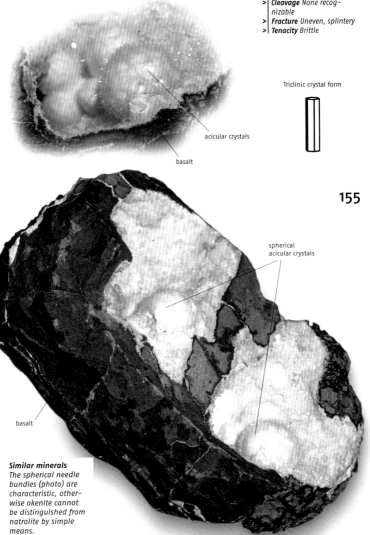

acicular crystals

basalt

Triclinic crystal form

155

spherical acicular crystals

basalt

Similar minerals
The spherical needle bundles (photo) are characteristic, otherwise okenite cannot be distinguished from natrolite by simple means.

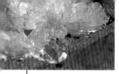

Hemimorphite

$Zn_4[(OH)_2/Si_2O_7] \cdot H_2O$

Occurrences In the oxidation zone of zinc deposits, where sufficient silica is present.

> **Hardness** 5
> **Density** 3.3–3.5
> **Luster** Vitreous
> **Cleavage** Perfect
> **Fracture** Conchoidal
> **Tenacity** Brittle

The white prismatic to tabular crystals of hemimorphite are colorless to white and, more rarely, yellowish, greenish, brown or even turquoise blue. The aggregates are radial, reniform, stalactitic and crusty.

prismatic crystals

Orthorhombic crystal form

Similar minerals Barite is much heavier; cerussite and anglesite have a different crystal form; aragonite effervesces in contrast to hemimorphite when dabbed with diluted hydrochloric acid.

156

tabular crystals

limonite

Apophyllite

$KCa_4[(F,OH)/(Si_4O_{10})_2] \cdot 8\ H_2O$

The colorless, white, yellow, green, brown or pink crystals are tabular, cubic, prismatic, also bipyramidal; the aggregates are flaky, granular, coarse. A typical feature of apophyllite is its pearly basal face.

pearly luster

prismatic crystal

Occurrences In bubble cavities of volcanic rocks, in druses and in fissures of ore passages, in alpine-type fissures.

> *Hardness* 4.5–5
> *Density* 2.3–2.4
> *Luster* Vitreous, pearly on the basal face
> *Cleavage* Perfect basal
> *Fracture* Uneven
> *Tenacity* Brittle

Similar minerals
The crystal form and the strong pearly luster on the basal face distinguish apophyllite from all other minerals of these parageneses.

Tetragonal crystal form

157

crystal clusters

stilbite

Scheelite
CaWO₄

$CaWO_4$

Occurrences *In pegmatites, pneumatolytic veins, hydrothermal gold ore veins, in alpine-type fissures.*

> **Hardness** 4.5–5
> **Density** 5.9–6.1
> **Luster** Greasy
> **Cleavage** Mostly poorly recognizable
> **Fracture** Conchoidal
> **Tenacity** Brittle

Tetragonal crystal form

Scheelite forms colorless, white, yellow, orange to red, mostly bipyramidal crystals, which rarely have a basal face. Often it exhibits coarse granular masses, which show a typical greasy luster. When exposed to UV light, scheelite glows intensely blue to yellow.

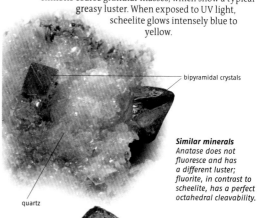

bipyramidal crystals

quartz

Similar minerals
Anatase does not fluoresce and has a different luster; fluorite, in contrast to scheelite, has a perfect octahedral cleavability.

158

greasy luster

bipyramidal

Apatite
Ca₅[(F,Cl)/(PO₄)₃]

Apatite forms colorless, white, yellow, blue, green to red prismatic, long to short columnar, tabular crystals, which sometimes appear spherical through many surfaces and which are raised and embedded. Aggregates can be acicular, radial, spherical, reniform, or coarse.

Occurrences *In all igneous rocks and their fissures and in cavities, in pegmatites, alpine-type fissures, in sediments.*

> ***Hardness*** 5
> ***Density*** 3.16–3.22
> ***Luster*** Vitreous
> ***Cleavage*** Sometimes discernible basal cleavage
> ***Fracture*** Conchoidal
> ***Tenacity*** Brittle

reniform carbonate apatite

Hexagonal crystal form

prismatic crystals

Similar minerals
Quartz, beryl and phenakite are harder; calcite, pyromorphite and mimetite are softer.

muscovite

Analcime
Na[AlSi₂O₆] · H₂O

$Na[AlSi_2O_6] \cdot H_2O$

Occurrences In bubble cavities of volcanic rocks, in ore veins.

> **Hardness** 5.5
> **Density** 2.2–2.3
> **Luster** Vitreous
> **Cleavage** Indiscernible
> **Fracture** Conchoidal
> **Tenacity** Brittle

Analcime forms almost only deltoidal icositetrahedron, rarely cubes with beveled edges. Its crystals are colorless to white, yellowish and reddish and mostly raised. Analcime is rarely crudely embedded.

deltoidal icositetrahedron

apophyllite

embedded crystals

Isometric crystal form

Similar minerals
Leucite in raised crystal form is indistinguishable by simple means; apophyllite has excellent cleavability, chabasite has a different crystal form.

160

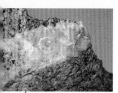

Pectolite
Ca₂NaH[Si₃O₉]

$Ca_2NaH[Si_3O_9]$

Occurrences Basic effusive rocks in fissures.

> **Hardness** 5
> **Density** 2.8
> **Luster** Vitreous, silky in aggregates
> **Cleavage** None
> **Fracture** Conchoidal, fibrous in aggregates
> **Tenacity** Brittle

Pectolite rarely forms white prismatic crystals. It often occurs as fibrous, radial aggregates. Blue radial aggregates are called larimar and are used under this name for jewelry.

blue larimar, polished

Triclinic crystal form

radial aggregate

Similar minerals
Wollastonite occurs in a completely different paragenesis, tobermorite is usually more fine-grained. The blue radial aggregates are very characteristic and should not be confused with any other precious or semi-precious stones.

Datolite

CaB[OH/SiO₄]

Datolite forms short prismatic to thick tabular crystals, from white to yellowish and greenish in color, more often granular, fibrous, reniform, coarse.

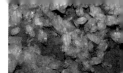

thick tabular crystals

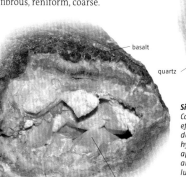

basalt

quartz

agate

free-growing crystals

Occurrences In bubble cavities of volcanic rocks, in ore veins, in alpine-type fissures, in boron-rich skarn deposits.

> **Hardness** 5–5.5
> **Density** 2.9–3
> **Luster** Vitreous, greasy on fracture surfaces
> **Cleavage** None
> **Fracture** Conchoidal
> **Tenacity** Brittle

Monoclinic crystal form

Similar minerals
Calcite is softer and effervesces when dabbed with diluted hydrochloric acid; apophyllite is softer and has a different luster and cleavability; danburite has a different cleavability.

Willemite

Zn₂[SiO₄]

Willemite forms short to long prismatic crystals and radial, reniform aggregates. Often it is grainy, coarse. The color varies from colorless, white to blue, green and yellow. Willemite fluoresces intensively yellow-green under UV light.

spherical crystal aggregate

cerussite

prismatic crystal

cerussite crystals

Occurrences In the oxidation zone of zinc deposits, in metamorphic zinc deposits.

> **Hardness** 5.5
> **Density** 4
> **Luster** Greasy vitreous
> **Cleavage** None
> **Fracture** Splintery
> **Tenacity** Brittle

Trigonal crystal system

Similar minerals
Calcite, pyromorphite, mimetite and vanidinite are softer, the same applies to cerussite; calcite effervesces when dabbed with diluted hydrochloric acid.

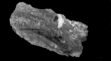

Titanite, Sphene

$CaTi[O/SiO_4]$

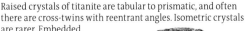

Occurrences In mag-
matites and crystalline
schists, raised crystals
in alpine-type fissures,
pegmatites, in marbles.

> *Hardness* 5–5.5
> *Density* 3.4–3.6
> *Luster* Resinous
> *Cleavage* Poorly recog-
> nizable
> *Fracture* Conchoidal
> *Tenacity* Brittle

Raised crystals of titanite are tabular to prismatic, and often
there are cross-twins with reentrant angles. Isometric crystals
are rarer. Embedded
crystals are usually
envelope-shaped. Colors
vary from yellow,
green to brown, pink
or blue.

chlorite

twin

re-entrant angle

Similar minerals
*Anatase is clearly
tetragonal, monazite
glows green when
exposed to unfiltered
UV light. Brookite is
typically thin tabular
or black (arcanite).*

Monoclinic crystal form

162

resinous

tabular crystal

Scolecite

$Ca[Al_2Si_3O_{10}] \cdot 3\ H_2O$

Scolecite crystals are acicular to prismatic, often in bundled to radial aggregates. Scolecite is almost always raised, rarely embedded.

prismatic crystals

acicular crystals

calcite

Similar minerals
Natrolite is generally somewhat more fine-grained and rather limited to volcanic rocks, but otherwise can hardly be distinguished from similar scolecite by simple means.

Occurrences *In bubble cavities of volcanic rocks, in druses and in fissures of ore passages, in alpine-type fissures.*

> **Hardness** 5.5
> **Density** 2.26–2.4
> **Luster** Vitreous
> **Cleavage** Perfect, but poorly recognizable due to the acicular crystals
> **Fracture** Conchoidal
> **Tenacity** Brittle

Monoclinic crystal form

Natrolite

$Na_2[Al_2Si_3O_{10}] \cdot 2\ H_2O$

Natrolite crystals are prismatic. Rarely do they have clearly visible end faces. Frequently, they are also long-prismatic to acicular. Even more frequently, however, they form radial to spherical aggregates or fibrous crusts. They are mostly raised, rarely embedded.

prismatic crystal

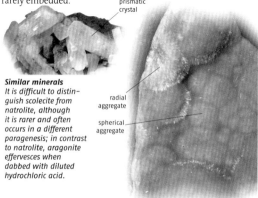

Similar minerals
It is difficult to distinguish scolecite from natrolite, although it is rarer and often occurs in a different paragenesis; in contrast to natrolite, aragonite effervesces when dabbed with diluted hydrochloric acid.

radial aggregate

spherical aggregate

Occurrences *In bubble cavities of volcanic rocks, in syenites and nepheline syenites.*

> **Hardness** 5–5.5
> **Density** 2.2–2.4
> **Luster** Vitreous
> **Cleavage** Perfect, but not recognizable due to fine-grained habit
> **Fracture** Conchoidal
> **Tenacity** Brittle

Orthorhombic crystal form

Tremolite, Grammatite
$Ca_2Mg_5[OH/Si_4O_{11}]_2$

Occurrences *In marbles, dolomites, talc schists.*

> ***Hardness*** *5.5–6*
> ***Density*** *2.9–3.1*
> ***Luster*** *Vitreous*
> ***Cleavage*** *Usually not recognizable due to radial habit*
> ***Fracture*** *Fibrous*
> ***Tenacity*** *Brittle*

Monoclinic crystal form

Tremolite forms white to green long prismatic crystals. It is often columnar, radial, and almost always embedded. Large prismatic crystals are very rare; tremolite usually forms fibrous, radial aggregates.

Similar minerals
In contrast to trem- olite, wollastonite is soluable in hydrochloric acid; actinolite is always distinctly green; strontianite is softer; natrolite and scol- ecite occur in a dif- ferent paragenesis; aragonite effervesces when dabbed with diluted hydrochloric acid.

tremolite colored pink by manganese

164

radial fibrous aggregate

silky luster

Childrenite

(Fe,Mn)Al[(OH)₂/PO₄] · H₂O.

$(Fe,Mn)Al[(OH)_2/PO_4] \cdot H_2O.$

Childrenite forms prismatic, long tabular crystals, radial aggregates. The crystal bundles are light brown to dark brown in color.

Occurrences As a young formation in druses in phosphate pegmatites.

> **Hardness** 5–5.5
> **Density** 3
> **Luster** Vitreous
> **Cleavage** Mostly unrecognizable
> **Fracture** Conchoidal
> **Tenacity** Brittle

prismatic crystals

quartz

long tabular crystal

Similar minerals
The manganese-rich end member phosphorite cannot be distinguished by simple means, however, both do not occur together; apatite crystals are distinctly six-sided.

Monoclinic crystal form

feldspar

Eosphorite

(Mn,Fe)AlPO₄(OH)₂ · H₂O

$(Mn,Fe)AlPO_4(OH)_2 \cdot H_2O$

Eosphorite crystals are prismatic, long tabular to acicular. Often as crystal bundles or radial aggregates. The crystals are colorless, yellowish, brown.

prismatic crystal

Occurrences In phosphate pegmatites, mostly raised in druses.

> **Hardness** 5–5.5
> **Density** 3
> **Luster** Vitreous
> **Cleavage** None
> **Fracture** Conchoidal
> **Tenacity** Brittle

acicular crystal bundles

Similar minerals
Eosphorite cannot be distinguished from childrenite by simple means. Apatite is always distinctly hexagonal; aragonite effervesces when dabbed with diluted hydrochloric acid.

Monoclinic crystal form

quartz

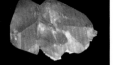

Brazilianite

NaAl₃[(OH)₂/PO₄]₂

Occurrences *In druses of pegmatites, as a conversion product of primary phosphates.*

> **Hardness** *5.5*
> **Density** *2.98*
> **Luster** *Vitreous*
> **Cleavage** *Perfect, perpendicular to the B-axis*
> **Fracture** *Uneven*
> **Tenacity** *Brittle*

Monoclinic crystal form

Brazilianite forms prismatic, long tabular to thick tabular crystals, mostly raised, rarely coarse and embedded. The crystals are often yellow in color, less often whitish.

raised crystal

muscovite

tabular crystal

Similar minerals
Topaz and albite are harder, the cleavage parallel to the longitudinal extension is very characteristic. Euclase is much harder; herderite has a different crystal form; calcite effervesces when dabbed with diluted hydrochloric acid.

Bavenite

Ca₄Al₂Be₂[(OH)₂/Si₃O₁₀/Si₆O₁₆]

Occurrences *In druses in pegmatites, in alpine-type fissures.*

> **Hardness** *5.5*
> **Density** *2.7*
> **Luster** *Vitreous, pearly on cleavage planes*
> **Cleavage** *Perfect*
> **Fracture** *Flaky*
> **Tenacity** *Brittle*

Orthorhombic crystal form

tabular crystal

Bavenite forms thick and thin tabular to acicular crystals, often grouped into rosettes. It is fibrous, flaky, powdery, coarse, raised and replaces beryl crystals.

spherical aggregates

radial aggregate

Similar minerals
Stilbite and laumonite have a different crystal form; tremolite is harder; the paragenesis of bavenite with other beryllium minerals is typical.

Turquoise

$CuAl_6[(OH)_2/PO_4]_4 \cdot H_2O$

Turquoise rarely forms tiny crystals, usually it is reniform, bulbous, coarse. It is turquoise blue, more rarely green, often with black veins. Turquoise is often used as a gemstone: Cabochons are used as ring stones and for brooches and pendants. Necklaces are made from baroque stones and beads.

limestone

turquoise vein

Occurrences *In the oxidation zone of copper deposits, as stringers and veins in phosphorus-rich slates.*

> ***Hardness*** *6*
> ***Density*** *2.91*
> ***Luster*** *Waxy to dull*
> ***Cleavage*** *None*
> ***Fracture*** *Uneven*
> ***Tenacity*** *Brittle*

Triclinic crystal form

calcite

turquoise

manganese oxides

Similar minerals
Color and paragenesis make turquoise unmistakable. Turquoise impregnated with synthetic resin shows a clear scratch mark and has a clear resin odor when scratched with a glowing needle, as does turquoise powder solidified with synthetic resin. Dyed magnesite is softer and discolors when dabbed with hydrochloric acid.

polished slab

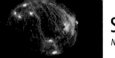

Sodalite

$Na_8[Cl_2/AlSiO_4)_6]$

Occurrences
*Rock-forming compo-
nent in magmatites, in
volcanic ejecta, crystals
in fissures.*

> **Hardness** 5–6
> **Density** 2.3
> **Luster** Vitreous, while its
> fracture is greasy
> **Cleavage** Mostly indis-
> cernible
> **Fracture** Conchoidal
> **Tenacity** Brittle

Sodalite rarely forms crystals (almost always rhombic
dodecahedron), usually it is massive, fibrous and coarse. It is
dark blue, greenish or colorless white. The blue variant is valued
as a decorative stone in architecture and as a gemstone.
Cabochons are used as ring stones or for brooches, pendants,
beads and necklaces.

Isometric crystal form

purple hackmanite
is a special variety of
sodalite

168

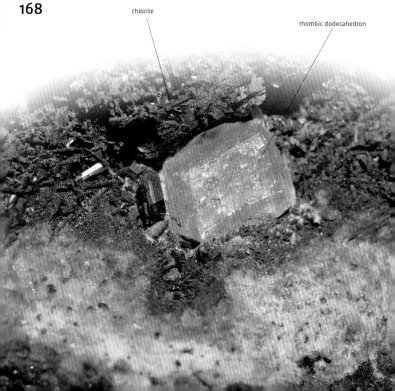

chlorite

rhombic dodecahedron

Hureaulite

$(Mn,Fe)_5H_2[PO_4]_4 \cdot 4\ H_2O$

Hureaulite crystals are prismatic, often with oblique end faces, and tabular. The aggregates are radial, coarse. Its color varies from colorless, pinkish red to yellowish pink to brown. Hureaulite almost always forms raised crystals.

Occurrences *In druses and cavities in phosphate pegmatites.*

> **Hardness** 5
> **Density** 3.2
> **Luster** *Vitreous*
> **Cleavage** *None*
> **Fracture** *Uneven*
> **Tenacity** *Brittle*

Monoclinic crystal form

hureaulite crystals in druse

169

vitreous luster

prismatic crystals

oblique end face

Similar minerals
Strengite and phosphosiderite have a different crystal form, just like apatite. Quartz and feldspar are harder than hureaulite.

Anatase
TiO$_2$

Occurrences *Raised in alpine-type fissures, embedded in clays and sandstones.*

> **Hardness** 5.5–6
> **Density** 3.8–3.9
> **Luster** *Metallic to adamantine*
> **Cleavage** *Mostly indiscernible*
> **Fracture** *Uneven*
> **Tenacity** *Brittle*

Tetragonal crystal form

Anatase forms pointed to flat bipyramids, prismatic and tabular crystals. Its crystals are almost always raised, often horizontally striated. The color is variable: colorless, pink, red, yellow, brown, blue, green, black.

bipyramidal crystal

prismatic crystal

Similar minerals
Magnetite and hematite have a black and red streak (respectively); brookite has a different crystal form; scheelite fluoresces intensively when exposed to ultraviolet light.

Brookite, Arkansite
TiO$_2$

Occurrences *In alpine-type fissures, in cavities in alkaline rocks.*

> **Hardness** 5.5–6
> **Density** 4.1
> **Luster** *Adamantine*
> **Cleavage** *Indiscernible*
> **Fracture** *Uneven*
> **Tenacity** *Brittle*

Orthorhombic crystal form

Brookite always occurs in raised crystals. There are two very different types: One consists of thin tabular crystals, often vertically striated and often with a dark hourglass pattern. The other is called arcanite, which is rarer and forms seemingly hexagonal bipyramids.

black arcanite crystals

hourglass drawing

albite

Similar minerals
Hematite has a different streak; anatase is always clearly tetragonal; the thin tabular crystals with hourglass pattern are unmistakable.

Eudialyte

$Na_4(Ca,Fe,Ce)_2ZrSi_6O_{17}(OH)_2$

Eudialyte forms yellowish brown, pink to deep red thick tabular, prismatic to isometric crystals, that are mostly embedded. It is often a coarse accessory mineral.

thick tabular crystal

coarse eudialyte

feldspar

Similar minerals
If the paragenesis and the typical color of eudialyte are taken into account, confusion is not possible; garnet is much harder; feldspar has a perfect cleavage; nepheline is usually a different color.

nepheline

Occurrences *In alkaline rocks, also as a rock-forming mineral.*

> **Hardness** 5–5.5
> **Density** 2.8
> **Luster** Vitreous to greasy
> **Cleavage** None
> **Fracture** Conchoidal
> **Tenacity** Brittle

Trigonal crystal system

Monazite

$CePO_4$

Monazite forms orange to brown thick tabular to prismatic crystals, which occur both raised and embedded. It is seldom coarse. It sometimes occurs as a heavy mineral in placers, where it is obtained industrially as a raw material for cerium. Crystals in alpine-type fissures are often transparent.

chlorite

embedded crystal

tabular crystal

Occurrences *In magmatites, large crystals in pegmatites, in placers, in alpine-type fissures.*

> **Hardness** 5–5.5
> **Density** 4.9–5.5
> **Luster** Vitreous to greasy
> **Cleavage** Sometimes visible
> **Fracture** Conchoidal
> **Tenacity** Brittle

Similar minerals
Titanite has a different crystal form; xenotime is clearly tetragonal; rutile has a very good cleavage and is more metallic; gadolinite has a greenish streak.

Monoclinic crystal form

Leucite
KAlSi₂O₆

KAlSi$_2$O$_6$

Occurrences *In volcanic rocks, basalts, tephrites, leucitites.*

> **Hardness** 5.5–6
> **Density** 2.5
> **Luster** Vitreous
> **Cleavage** None
> **Fracture** Uneven
> **Tenacity** Brittle

The colorless to mostly white crystals of leucite are pseudocubic. The typical deltoidicositetrahedra are almost always embedded. Sometimes they are transformed into feldspar or clay minerals while retaining their crystal form.

leucitite, a rock made with leucite as its primary constituent

Tetragonal crystal form

Similar minerals
Analcime is mostly raised, but otherwise difficult to distinguish; nepheline has a different crystal form; sanidine shows perfect cleavability.

deltoidal icositetrahedron

tephrite

Diopside

$CaMg[Si_2O_6]$

Diopside forms colorless, white, yellow, brown, but mostly green crystals. The crystals are prismatic. The aggregates are radial, columnar. As a rock-forming mineral, diopside is coarse. Transparent crystals are seldom used for jewelry.

Occurrences In plutonic rocks, marbles, calc-silicate rocks, in alpine-type fissures.

> *Hardness* 6
> *Density* 3.3
> *Luster* Vitreous
> *Cleavage* Recognizable, cleavage angle approx. 90°
> *Fracture* Uneven
> *Tenacity* Brittle

Similar minerals Hornblende has a different cleavage angle; epidote has a different crystal form and a very typical color; karpolite is always characteristically radial and significantly softer.

Monoclinic crystal form

parallelly grown crystals

prismatic crystal

epidote

Wagnerite
$(Mg,Fe)_2(PO_4)F$

Occurrences *In meta-morphic rocks, in pegmatites.*

> ***Hardness*** *5*
> ***Density*** *3.15*
> ***Luster*** *Vitreous to resinous*
> ***Cleavage*** *None recognizable*
> ***Fracture*** *Uneven, splintery*
> ***Tenacity*** *Brittle*

Wagnerite forms yellowish, honey-colored, more rarely orange and gray crystals. They are prismatic, elongated, longitudinally striated and sometimes form columnar aggregates. Wagnerite is often also massive and coarse.

broken and reconsolidated single crystal

fracture point

Monoclinic crystal form

174

prismatic crystal

adjoining rock

Similar minerals
Quartz is harder, gypsum is softer. Siderite has a perfect rhombohedral cleavability; aragonite effervesces when dabbed with diluted hydrochloric acid.

Potassium Feldspar, Orthoclase, Microcline

$K[AlSi_3O_8]$

The crystals of potassium feldspar are prismatic, thick and thin tabular (sanidine), also rhombohedral (adularia). It often forms twins with re-entrant angles as well as raised crystals in pegmatites (microcline). However, it occurs most frequently in large, coarse masses. The color varies from colorless to white, yellow, brown, green to flesh red. Some variants are used for jewelry: green (amazonite), iridescent orange (sunstone), white, partly milky with a bluish glow (moonstone).

Occurrences *In igneous and metamorphic rocks, pegmatites, in alpine-type fissures, in ore veins.*

> ***Hardness*** 6
> ***Density*** 2.53–2.56
> ***Luster*** *Vitreous*
> ***Cleavage*** *Perfect*
> ***Fracture*** *Conchoidal*
> ***Tenacity*** *Brittle*

Monoclinic crystal form (sanidine and orthoclase) and triclinic crystal form (microcline)

Similar minerals
Quartz has no cleavability; calcite and fluorite are much softer. Plagioclase is not easy to distinguish by simple means.

amazonite crystal

175

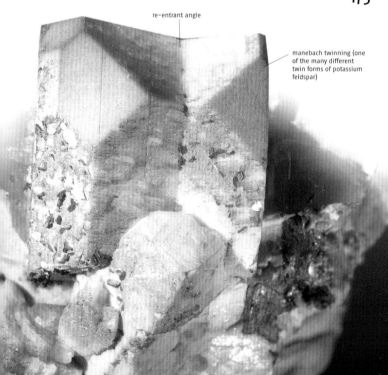

re-entrant angle

manebach twinning (one of the many different twin forms of potassium feldspar)

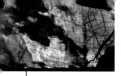

Plagioclase
(Na,Ca)[(Al,Si)$_2$Si$_2$O$_8$]

Occurrences *In igneous and metamorphic rocks, pegmatites, in alpine-type fissures, in ore veins.*

> **Hardness** 6–6.5
> **Density** 2.61–2.77
> **Luster** Vitreous
> **Cleavage** Perfect, 90° cleavage angle
> **Fracture** Conchoidal
> **Tenacity** Brittle

Triclinic crystal form

Plagioclases form a solid solution series with the two end members: albite Na[AlSi$_3$O$_8$] and anorthite Ca[Al$_2$Si$_2$O$_8$]. The intermediate members have different names depending on the mixing ratio:

Oligoclase 70–90% albite
Andesine   50–70% albite
Labradorite 30–50% albite
Bytownite   10–30% albite

They are white, greenish, reddish and form prismatic to tabular crystals which are often twinned. Plagioclase with a blue to colorful iridescence on dark background are polished as labradorite or spectrolite.

pericline (a special twi... form of albite found i... alpine-type fissures)

Similar minerals
Quartz has no cleavability; calcite, barite, gypsum and dolomite are softer; potassium feldspar shows other crystal forms.

176

tabular crystal

vitreous luster

Rhodonite
CaMn₄[Si₅O₁₅]

Rhodonite forms tabular (often sharp to prismatic crystals, though more often it is fibrous, coarse. Deep red solid rhodonite with black veins is used for jewelry.

Occurrences *In meta-morphic manganese deposits, ore veins.*

> **Hardness** 5.5–6.5
> **Density** 3.73
> **Luster** Vitreous
> **Cleavage** Perfect
> **Fracture** Uneven
> **Tenacity** Brittle

dark veins caused by manganese oxides

cabochon

Triclinic crystal form

Similar minerals
Rhodochrosite is softer. The color and the typical black vein pattern means that rhodonite can't be confused with any other gemstone.

prismatic crystal

siderite

Opal
$SiO_2 \cdot n\ H_2O$

Occurrences In
cavities of volcanic
rocks, in sediments at
groundwater level, as a
deposit found around
hot springs (geyserite).

> **Hardness** 5–6.5
> **Density** 1.9–2.2
> **Luster** Waxy to vitreous,
>   sometimes iridescent play
>   of colors
> **Cleavage** None
> **Fracture** Conchoidal
> **Tenacity** Brittle

Opal is always amorphous, coarsely embedded. It can be found
as a filling in druses and as reniform, spherical, drop-shaped
aggregates. It displays a wide variety of colors, from white (milky
opal), red-orange (fire opal), brown, green to colorless (hyalite)
and bluish to blackish with a clear play of colors. The latter is
classified as noble opal and is a valuable gemstone.

178

Similar minerals
Chalcedony can be
similarly formed,
which makes it
difficult to be dis-
tinguished by simple
means; noble opal
always differs by its
play of colors.

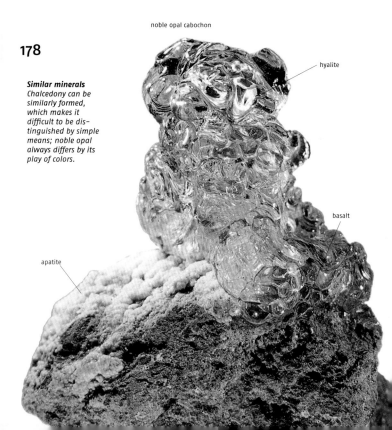

noble opal

limonite

noble opal cabochon

hyalite

basalt

apatite

Cassiterite, Tinstone
SnO_2

Cassiterite forms prismatic to acicular crystals, knee-shaped twins and radial aggregates ("wood tin"). It is often also coarse and usually brown. More rarely it is also colorless, reddish or almost black.

Occurrences In pegmatites, pneumatolytic veins and displacements, hydrothermal veins, placers.

> **Hardness** 7
> **Density** 6.8–7.1
> **Luster** Pearlescent to greasy
> **Cleavage** Hardly visible
> **Fracture** Conchoidal
> **Tenacity** Brittle

greasy luster

Similar minerals
Crystal form and high density distinguish cassiterite from almost all other minerals; rutile is lighter or, if dark, clearly more metallic and then also has a more distinct cleavage.

cassiterite gravel, so-called wood tin

Tetragonal crystal form

spodumene

twin aggregate

albite crystals

Jadeite
NaAl[Si$_2$O$_6$]

Occurrences *In crystal-line schist.*

> **Hardness** *6.5*
> **Density** *3.2–3.3*
> **Luster** *Vitreous*
> **Cleavage** *Hardly recognizable due to the dense habit*
> **Fracture** *Conchoidal*
> **Tenacity** *Tough*

Jadeite rarely forms short prismatic crystals, rather they are mostly granular, fibrous, dense aggregates of white, yellow, green and purple color. This material is used as jade for jewelry and for the production of handcrafted items.

Monoclinic crystal form

Similar minerals
Nephrite is somewhat softer, but difficult to distinguish from jadeite. The former is mostly yellowish green; grossular (Transvaal jade) is darker green than jadeite; serpentine is much softer.

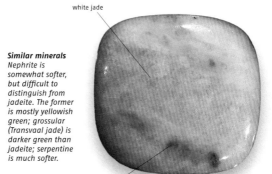

white jade

green jade

cabochon

180

polished surface

purple jade

Vesuvianite, Idocrase

$Ca_{10}(Mg,Fe)_2Al_4[(OH)_4/(SiO_4)_5/(Si_2O_7)_2]$

Vesuvianite forms long to short prismatic crystals, mostly columnar but also tabular, or radial aggregates and is often also granular, coarse. Its color varies from yellow and brown to green and purple.

short prismatic crystal

blue calcite

purple vesuvianite

Similar minerals
Grossular is cubic, but often difficult to distinguish from short-prismatic vesuvianite; zircon is heavier and harder; chondrodite, humite and clinohumite have a different crystal form.

Occurrences In metamorphic limestones, in alpine-type fissures, in volcanic ejecta.

> **Hardness** 6.5
> **Density** 3.27–3.45
> **Luster** Vitreous to greasy
> **Cleavage** None
> **Fracture** Conchoidal
> **Tenacity** Brittle

Tetragonal crystal form

Wiluite

$Ca_{19}(Al,Mg,Fe,Ti)_{13}(B,Al)_5Si_{18}O_{68}(O,OH)_{10}$

Wiluite is the boron-containing variety of vesuvianite. It forms brown to brownish-green long to short-prismatic crystals, which are mostly embedded. Less frequently, it is also granular, coarse.

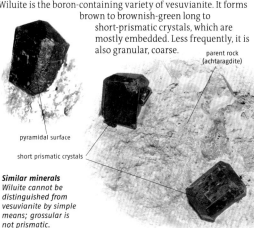

parent rock (achtaragdite)

pyramidal surface

short prismatic crystals

Similar minerals
Wiluite cannot be distinguished from vesuvianite by simple means; grossular is not prismatic.

Occurrences In serpentine skarn.

> **Hardness** 6.5
> **Density** 3.27–3.45
> **Luster** Vitreous to greasy
> **Cleavage** None
> **Fracture** Conchoidal
> **Tenacity** Brittle

Tetragonal crystal form

Sugilite
$(K,Na)(Na,Fe)_2(Li_2Fe)[Si_{12}O_{30}]$

Occurrences In meta-morphic manganese deposits.

> **Hardness** 6–7
> **Density** 2.74
> **Luster** Vitreous to dull
> **Cleavage** None recognizable
> **Fracture** Conchoidal
> **Tenacity** Tough

Sugilite rarely shows prismatic, vertically striated crystals. It generally forms purple dense masses, which are granular, coarse. Sugilite is used to make jewelry, dense masses are also used to make handcrafted items.

Hexagonal crystal form

Similar minerals
The rare charoite from Siberia is always clear, easily recognizable with the naked eye, fibrous and not granular like the sugilite. The former is also usually more distinctly blue-purple. Otherwise, the color of sugilite is extraordinarily characteristic and unmistakable.

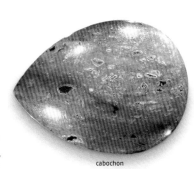

cabochon

182

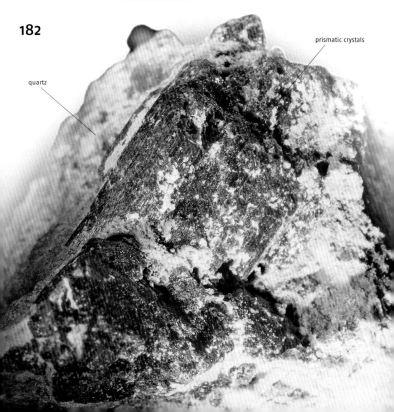

quartz

prismatic crystals

Prehnite

Ca₂Al[(OH)₂/AlSi₃O₁₀]

Prehnite forms colorless, white, often green reniform to
spherical aggregates and sometimes also stalactitic formations.
More rare are tabular to prismatic crystals.

Occurrences In bubble
cavities of volcanic
rocks, in druses in
pegmatites, in alpine-
type fissures.

> **Hardness** 6–6.5
> **Density** 2.8–3
> **Luster** Vitreous
> **Cleavage** Recognizable
> basal cleavage
> **Fracture** Uneven
> **Tenacity** Brittle

prismatic
crystals

Similar minerals
*Wavellite occurs in
a different para-
genesis and crystal
form; stilbite and
heulandite have a
different hardness,
crystal form and
much better cleav-
ability.*

Orthorhombic crystal form

parquet-like
surface

spherical
aggregate

Cordierite, Dichroite

$Mg_2Al_3[AlSi_5O_{18}]$

Cordierite is a rock-forming mineral that is mostly coarse, more
rarely are embedded six- to twelve-sided prisms. The color varies
from gray to blue, purple, yellow to greenish. Viewed from
different angles, cordierite shows different colors from almost
colorless to purple. This is called pleochroism.

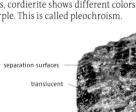

separation surfaces

translucent

Orthorhombic crystal form

Similar minerals
Tourmaline is glossy
black; it is not always
possible to distinguish
coarse cordierite from
quartz by simple
means.

184

pyrrhotine

prismatic crystal

Axinite

$Ca_2(Fe,Mg,Mn)Al_2[OH/BO_3/Si_4O_{12}]$

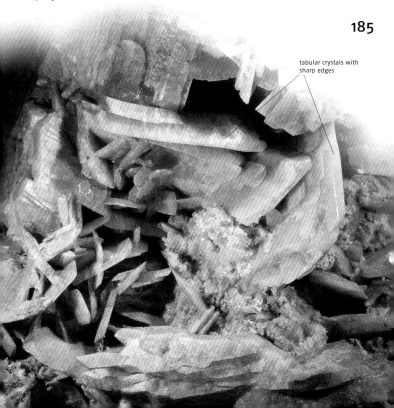

The tabular crystals of axinite are usually very sharp-edged and almost always raised. More rarely, axinite is also columnar, fibrous and coarse. Its color varies from brown, gray, blue, greenish to purple.

Occurrences In calc-silicate rocks, contact metasomatic deposits, in alpine-type fissures, in druses in pegmatites.

> **Hardness** 6.5–7
> **Density** 3.3
> **Luster** Vitreous
> **Cleavage** Poorly visible
> **Fracture** Conchoidal
> **Tenacity** Brittle

Similar minerals
The sharp-edged crystals of axinite are unmistakable; rhodonite is always more or less red; adular and albite never have such sharp edges.

raised tabular crystal

Triclinic crystal form

tabular crystals with sharp edges

Zoisite
$Ca_2Al_3[O/OH/SiO_4/Si_2O_7]$

Occurrences *In metamorphic rocks, especially blue slate facies, in pegmatites.*

> **Hardness** *6–7*
> **Density** *3.15–3.36*
> **Luster** *Vitreous*
> **Cleavage** *Perfect*
> **Tenacity** *Brittle*
> **Fracture** *Uneven*

Zoisite forms long to short prismatic or long tabular crystals, but mostly radial, columnar aggregates. It is gray, light brown, gray-brown and blue. The latter of which is used as a gemstone under the name tanzanite.

blue tanzanite

feldspar

columnar crystals

Orthorhombic crystal form

Similar minerals
Clinozoisite is not always easy to distinguish from zoisite; actinolite and hornblende have a perfect cleavage angle of 120°.

Clinozoisite
$Ca_2Al_3[O/OH/SiO_4/Si_2O_7]$

Occurrences *In hydrothermal veins and metamorphic rocks.*

> **Hardness** *6–7*
> **Density** *3.3–3.5*
> **Luster** *Vitreous*
> **Cleavage** *Perfect*
> **Tenacity** *Brittle*
> **Fracture** *Uneven*

Clinozoisite crystals are long tabular to prismatic, often radial and columnar, and rarely fibrous aggregates. Colors are whitish, gray, light brown and gray brown.

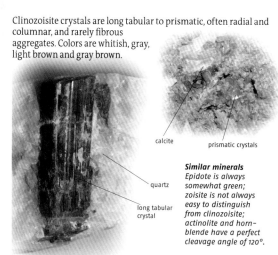

calcite

prismatic crystals

quartz

long tabular crystal

Monoclinic crystal form

Similar minerals
Epidote is always somewhat green; zoisite is not always easy to distinguish from clinozoisite; actinolite and hornblende have a perfect cleavage angle of 120°.

Almandine

$Fe_3Al_2[SiO_4]_3$

Almandine belongs to the garnet group. It forms red or reddish brown to brown crystals, often rhombic dodecahedrons, more rarely deltoidal icositetrahedrons. The crystals are almost always embedded. Red transparent variants are cut as precious stones.

Occurrences *In mica schists, gneisses, granulites, more rarely in pegmatites.*

> ***Hardness*** *6.5–7.5*
> ***Density*** *4.32*
> ***Luster*** *Vitreous*
> ***Cleavage*** *None*
> ***Fracture*** *Conchoidal*
> ***Tenacity*** *Brittle*

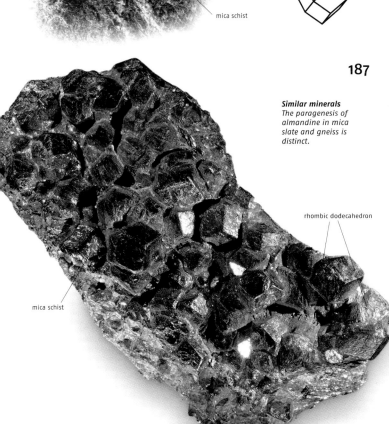

rhombic dodecahedron

mica schist

Isometric crystal form

Similar minerals
The paragenesis of almandine in mica slate and gneiss is distinct.

rhombic dodecahedron

mica schist

Pyrope
$Mg_3Al_2[SiO_4]_3$

Occurrences In ultraba-
sites, serpentinites and
placers.

> **Hardness** 7–7.5
> **Density** 3.58
> **Cleavage** None
> **Fracture** Conchoidal
> **Tenacity** Brittle

Pyrope belongs to the garnet group. Its intensively red crystals
are always embedded, rarely as well-formed rhombic dodecahedrons or deltoid icositetrahedrons but often as round grains and
in placers. Pyrope is often cut as a gemstone.

faceted pyrope

Isometric crystal form

inclusions of
rutile needles

peridotite

188

embedded
pyrope

Similar minerals
The paragenesis of
pyrope is distinct;
almandine is always
somewhat more
brownish, never purely
dark red.

Grossular

Ca₃Al₂[SiO₄]₃

$Ca_3Al_2[SiO_4]_3$

Grossular belongs to the garnet group. It shows well-formed rhombic dodecahedrons, deltoidal icositetrahedrons that are frequently raised. In addition, granular or coarse to dense masses appear. The color varies from colorless, yellow, yellow-brown, green to red (hessonite). Colorful dense masses are polished into cabochons and art objects.

Similar minerals
The paragenesis of grossular is very char-acteristic; vesuvianite is usually distinctly prismatic.

polished slab

red-green dense grossular

Occurrences *In contact marbles, in fissures found in serpentinites and rodingites.*

> ***Hardness*** 6.5–7
> ***Density*** 3.59
> ***Luster*** Vitreous
> ***Cleavage*** None
> ***Fracture*** Conchoidal
> ***Tenacity*** Brittle

Isometric crystal form

rhombic dodecahedron surface

Andradite
$Ca_3Fe_2[SiO_4]_3$

Occurrences In meta-morphic deposits, in fissures found in serpentines in volcanic rocks.

> *Hardness* 6.5–7.5
> *Density* 3.86
> *Luster* Vitreous
> *Cleavage* None
> *Fracture* Conchoidal
> *Tenacity* Brittle

Isometric crystal form

Andradite belongs to the garnet group. It shows well-formed rhombic dodecahedron, deltoidal icositetrahedron, frequently raised, but also granular coarse to dense masses. The color varies from colorless, yellow, brown, green to black (melanite). Transparent yellow and yellow-green crystals are also called demantoids or topazolites and are used to make jewelry.

demantoid

brown rhombic dodecahedron

serpentine

190

Similar minerals Grossular is often indistinguishable from andradite by simple means.

hedenbergite

Spessartite

$Mn_3Al_2[SiO_4]_3$

Spessartite belongs to the garnet group. It shows well-formed rhombic dodecahedron, deltoidal icositetrahedron, frequently raised. It also forms coarse masses in metamorphic manganese deposits. Colors vary from pink and orange to light to dark brown and red. Clear transparent orange crystals are used for making jewelry.

Occurrences *In metamorphic manganese deposits, in pegmatites and granites.*

> ***Hardness*** 7
> ***Density*** 4.19
> ***Luster*** Vitreous
> ***Cleavage*** None
> ***Fracture*** Conchoidal
> ***Tenacity*** Brittle

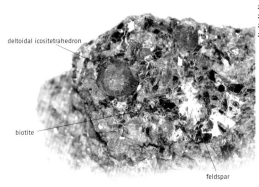

deltoidal icositetrahedron

biotite

feldspar

Isometric crystal form

galena

spessartine crystals embedded in galena

rhombic dodecahedron

Similar minerals
Almandine is more reddish brown, in contrast to spessartine it often exhibits rhombic dodecahedron.

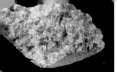

Olivine, Peridot, Chrysolite
$(Mg,Fe)_2[SiO_4]$

Occurrences Rock-forming component in magmatites, raised crystals in fissures, crystals and grains in crystalline limestones, in meteorites.

Olivine is an important rock-forming mineral. Thick tabular to prismatic crystals are rather rare. Transparent olivine crystals are cut to gemstones (known as peridot). Colors vary from yellowish green to bottle green, rarely olivine is brownish to red.

> **Hardness** 6.5–7
> **Density** 3.27–4.2
> **Luster** Vitreous, slightly greasy
> **Cleavage** Hardly recognizable
> **Fracture** Conchoidal
> **Tenacity** Brittle

Orthorhombic crystal form

Similar minerals
Apatite is softer, as is serpentine; beryl is harder and always has a six-sided cross-section.

faceted peridot

192

tabular crystal

striations

conchoidal fracture

Bertrandite

Be₄[(OH)₂/Si₂O₇]

Bertrandite forms colorless, white
and more rarely, yellowish tabular
crystals. The v-shaped twins are
almost always raised.

tabular crystal

colored by iron

long tabular crystals

quartz

Occurrences *In druses
in pegmatites, in
alpine-type fissures,
in pneumatolytic
deposits.*

> **Hardness** 6.5–7
> **Density** 2.6
> **Luster** Vitreous, pearly on
the basal face
> **Cleavage** Perfect basal
> **Fracture** Conchoidal
> **Tenacity** Brittle

Similar minerals
*Albite has a different
crystal form; barite
and muscovite are
much softer; tabular
quartz crystals are
sometimes difficult
to distinguish from
bertrandite.*

Orthorhombic crystal form

Staurolite

(Fe,Mg,Zn)₂Al₉[O₆/(OH)₂/(SiO₄)₄]

long tabular
crystal

Staurolite forms reddish-brown to
black-brown prismatic to tabular
crystals, which are often intergrown
to form typical cross-shaped twins
(right-angled or at about 60°).

disthene

Occurrences *Embedded
in mica schist and
gneisses.*

> **Hardness** 7–7.5
> **Density** 3.7–3.8
> **Luster** Vitreous
> **Cleavage** Hardly visible
> **Fracture** Conchoidal
> **Tenacity** Brittle

mica schist

cross twinning

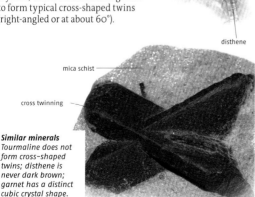

Similar minerals
*Tourmaline does not
form cross-shaped
twins; disthene is
never dark brown;
garnet has a distinct
cubic crystal shape.*

Monoclinic crystal form

Spodumene, Kunzite, Hiddenite

$LiAl[Si_2O_6]$

Occurrences Radial,
tabular embedded
crystals in pegma-
tites, raised crystals in
druses.

> **Hardness** 6.5–7
> **Density** 3.1–3.2
> **Luster** Vitreous
> **Cleavage** Perfect prismatic
> **Fracture** Splintery
> **Tenacity** Brittle

Spodumene forms tabular crystals that can grow to several
meters in size and are white and opaque. The colored varieties,
kunzite (pink) and hiddenite (green), are used for jewelry.

faceted kunzite

Monoclinic crystal form

194

tabular hiddenite crystal

tabular kunzite crystal

striations

Similar minerals
Feldspar has a differ-
ent cleavability; quartz
has no cleavability at
all; pink beryl has a
completely different
six-sided crystal form;
apatite is softer and
has a different six-
sided crystal form.

Quartz (Rock Crystal)
SiO₂

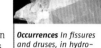

The colorless variety of quartz is called rock crystal. It is found in alpine-type fissures, in ore veins, in cavities of volcanic rocks, as a doubly terminated crystal in limestone and salt rocks, in cavities of pegmatites, and on fissures of quartz-rich rocks. Due to inclusions, opaque white quartz is called milky quartz. It is often a gangue mineral in hydrothermal veins. In fissures, it can also form beautiful crystals. Clear rock crystal is faceted and made into jewelry.

Occurrences In fissures and druses, in hydro-thermal veins.

> *Hardness 7*
> *Density 2.65*
> *Luster Vitreous to greasy*
> *Cleavage None*
> *Fracture Conchoidal*
> *Tenacity Brittle*

Japanese twins

Trigonal crystal form
(low quartz)

rock crystal

coarse quartz

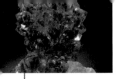

Quartz (Amethyst)
SiO₂

Occurrences *In druses in volcanic rocks, in hydrothermal veins, in alpine-type fissures.*

> **Hardness** *7*
> **Density** *2.65*
> **Luster** *Vitreous to greasy*
> **Cleavage** *None*
> **Fracture** *Conchoidal*
> **Tenacity** *Brittle*

The purple variety of quartz is called amethyst. It is primarily found in ore veins (here as long prismatic crystals) and in cavities of volcanic rocks. There, amethyst can form regular druses — cavities in which all walls are occupied by short prismatic amethyst crystals. If amethyst is heated, its color changes to brown-yellow. The result is artificial citrine. Clear amethyst is faceted and made into jewelry.

Trigonal crystal form
(low quartz)

faceted amethyst

striations

long prismatic crystal

typical purple color

Quartz (Smoky Quartz)
SiO₂

Smoky quartz forms smoky brown transparent to opaque deep black crystals. The latter are also called morion. Smoky quartz can be found primarily in pegmatites, in alpine-type fissures and in hydrothermal veins. The black color comes from natural radioactive irradiation, but there are also quartz crystals that are colored deep black by inclusions of bituminous organic substances. Clear smoky quartz is faceted and made into jewelry.

Occurrences *In alpine-type fissures, in pegmatites, in hydro- thermal veins.*

> **Hardness** 7
> **Density** 2.65
> **Luster** Vitreous to greasy
> **Cleavage** None
> **Fracture** Conchoidal
> **Tenacity** Brittle

faceted smoky quartz

transparent crystal

Trigonal crystal form
(low quartz)

197

opaque crystal

feldspar

Quartz (Rose Quartz/Citrine)
SiO₂

> *Hardness* 7
> *Density* 2.65
> *Luster* Vitreous to greasy
> *Cleavage* None
> *Fracture* Conchoidal
> *Tenacity* Brittle

Pink quartz, which only occurs in pegmatites, is called rose quartz. Well-formed rose quartz crystals seldom occur in druses. Yellow quartz crystals (citrine) are not very common in pegmatites. Quartz is a component of plutonic rocks, volcanic rocks, sedimentary rocks and metamorphic rocks. Beautiful crystals can be found in druses in pegmatites, in pneumatolytic veins, in ore veins, in hydrothermal quartz veins, in alpine-type fissures and in cavities in marble.

coarse rose quartz from pegmatite

Trigonal crystal form
(low quartz)

198

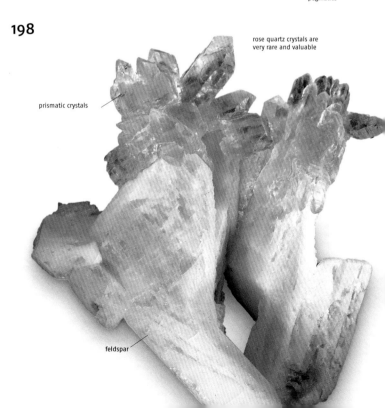

rose quartz crystals are very rare and valuable

prismatic crystals

feldspar

Quartz (Chalcedony)
SiO₂

Chalcedony is microcrystalline quartz, i.e. it forms the smallest crystals visible only under the microscope and which occur only as reniform aggregates, cavity fillings or stalactitic formations. It is colorless or multicolored: Chalcedony in the narrower sense (colorless, white, gray, blue, monochrome and striped), carnelian (translucent red to reddish brown), chrysoprase (green through inclusions of nickel minerals), onyx (alternating black and white layers).

Occurrences In hydro-thermal veins, volcanic rocks, in sedimentary rocks.

> *Hardness* 7
> *Density* 2.65
> *Luster* Vitreous to greasy
> *Cleavage* None
> *Fracture* Conchoidal to uneven
> *Tenacity* Brittle

Trigonal crystal form (low quartz)

basalt

reniform chalcedony

chalcedony in the form of fluorite crystals (pseudomorphism)

greasy luster

199

Similar minerals
Fluorite and calcite are much softer.

Quartz (Jasper)
SiO$_2$

Occurrences In hydro-
thermal veins, volcanic
rocks, in sedimentary
rocks.

> **Hardness** 7
> **Density** 2.65
> **Luster** Vitreous to greasy
> **Cleavage** None
> **Fracture** Conchoidal to
> uneven
> **Tenacity** Brittle

Jasper is opaque microcrystalline quartz, which may have
different colors due to inclusions of other minerals (e.g.
hematite, limonite, chlorite). There's red, yellow, and green
jasper. The so-called heliotrope is green with red spots
(hematite). Flint is gray, brownish to almost black and is found in
nodules in sedimentary rocks.

Trigonal crystal form
(low quartz)

Similar minerals
*Fluorite and calcite
are significantly softer
and show excellent
cleavage properties.*

green jasper is also
called plasma

conchoidal fracture

200

chrysoprase is characterized by
inclusions of green-colored nickel
minerals

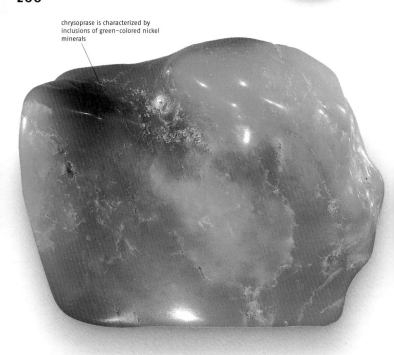

Quartz (Agate)
SiO_2

Agate is microcrystalline quartz and therefore does not form visible crystals, only reniform aggregates. It forms fillings in cavities as well as stalactite formations. The different agates are named after the patterns they show when cutting the nodules. Beautiful agates are processed into jewelry and handcrafted items.

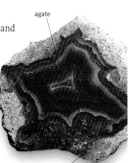

agate

surrounding rock

Occurrences *In hydrothermal veins, in fossilized woods, in cavities found in volcanic rocks.*

> **Hardness** 7
> **Density** 2.65
> **Luster** *Vitreous to greasy*
> **Cleavage** *None*
> **Fracture** *Conchoidal to uneven*
> **Tenacity** *Brittle*

Trigonal crystal form
(low quartz)

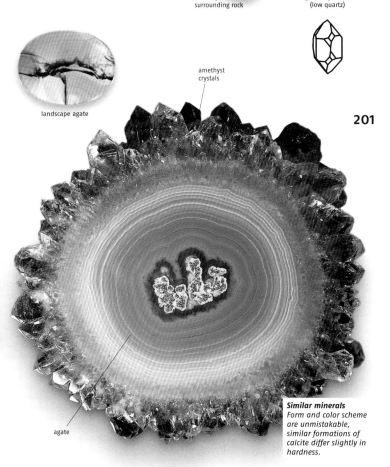

landscape agate

amethyst crystals

agate

Similar minerals
Form and color scheme are unmistakable, similar formations of calcite differ slightly in hardness.

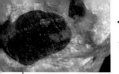

Tourmaline

The tourmalines are a group of related minerals.

> **Hardness** 7
> **Density** 3–3.25
> **Luster** Vitreous
> **Cleavage** None
> **Fracture** Conchoidal
> **Tenacity** Brittle

Schorl forms black tabular, prismatic to acicular crystals that are raised and embedded. Dravite is similarly formed, but more brown.

There are the following mixed members:
Elbaite $Na(Li,Al)_3Al_6[OH]_4/(BO_3)_3/Si_6O_{18}$
Dravite $NaMg_3(Al,Fe^{3+})_3Al_6[OH]_4/(BO_3)_3/Si_6O_{18}$
Schorl $NaFe_3^{2+}(Al,Fe)_6[OH]_4/(BO_3)_3/Si_6O_{18}$
Buergerite $NaFe_3^{3+}Al_6[F/O_3/(BO_3)_3/Si_6O_{18}$
Tsilaisite $NaMn_3Al_6[OH]_4/(BO_3)_3/Si_6O_{18}$
Uvite $CaMg_3(Al_5Mg)[OH]_4/(BO_3)_3/Si_6O_{18}$
Liddicoatite $Ca(Li,Al)_3Al_6[OH]_4/(BO_3)_3/Si_6O_{18}$

Trigonal crystal system

202

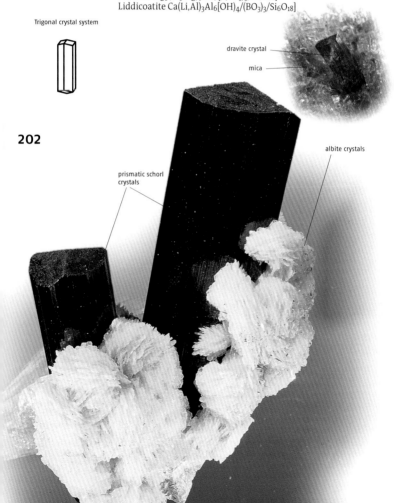

dravite crystal

mica

albite crystals

prismatic schorl crystals

Elbaite forms multicolored, prismatic to acicular crystals that are raised and embedded. There are the following color variants: rubellite (pink, red), verdelite (green), indigolite (blue). There are even multicolored elbaites. Transparent crystals are cut to gemstones.

Occurrences In granites, pegmatites, pneumatolytic veins, hydrothermal veins.

> **Hardness** 7
> **Density** 3–3.25
> **Luster** Vitreous
> **Cleavage** None
> **Fracture** Conchoidal
> **Tenacity** Brittle

Trigonal crystal system

203

typical triangular cross-section

bicolored elbaite, faceted

multicolored elbaite crystal

rubellite crystal

tabular albite crystals

albite crystals

Similar minerals
The generally distinct three-sided cross-section distinguishes tourmaline from all other minerals.

Boracite
Mg₃[Cl/B₇O₁₃]

Occurrences *Embedded in anhydrite or gypsum in salt deposits.*

> **Hardness** *7*
> **Density** *2.9–3*
> **Luster** *Vitreous*
> **Cleavage** *None*
> **Fracture** *Conchoidal*
> **Tenacity** *Brittle*

Boracite forms cubic, tetrahedral crystals, which are usually embedded in gypsum; in fibrous or dense formation it is called strassfurtite, named for where it is found. Its color varies from colorless, white, yellowish, greenish to bluish.

cubic crystal

strassfurtite

gypsum

Isometric crystal form

Similar minerals
Rock salt and fluorite are much softer and have good cleavage properties.

204

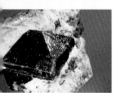

Zircon
Zr[SiO₄]

Occurrences *In magmatites, volcanic ejecta, in placers, pegmatites, in alpine-type fissures.*

> **Hardness** *7.5*
> **Density** *4.55–4.67*
> **Luster** *Adamantine-like, greasy on fractured surfaces*
> **Cleavage** *Hardly noticeable*
> **Fracture** *Conchoidal*
> **Tenacity** *Brittle*

Zircon always forms prismatic to bipyramidal crystals. It is raised but more often embedded and is practically never coarse. Its color varies from colorless to white, pink, yellow, green to blue and brown. Transparent crystals are cut into jewelry.

prismatic crystal

phlogopite

faceted zircon

Tetragonal crystal form

Similar minerals
Vesuvianite is softer; cassiterite is heavier. White and light blue sapphire don't have high birefringence like zircon; diamond has a much higher brilliancy.

Andalusite, Chiastolite

$Al_2[O/SiO_4]$

Andalusite forms thick columnar crystals with an almost square cross-section, as well as radial aggregates. It's almost always embedded. Some andalusites are called chiastolites because their cross-section shows a cross-shaped pattern similar to the Greek letter chi. Such stones are also used for jewelry.

prismatic crystal

chiastolite, polished cross-section

quartz

Occurrences In gneisses and mica schists, in slates and pegmatites.

> **Hardness** 7.5
> **Density** 3.1–3.2
> **Luster** Vitreous, but mostly cloudy
> **Cleavage** Mostly indiscernible
> **Fracture** Uneven
> **Tenacity** Brittle

Orthorhombic crystal form

Sillimanite

$Al_2[O/SiO_4]$

Sillimanite rarely forms prismatic to acicular individual crystals. Very often, the fibrous, radial, columnar aggregates are white. It's almost always embedded.

prismatic crystals

fibrous silica is called sillimanite, which is densely intergrown with quartz

quartz

Occurrences In magmatites, volcanic ejecta.

> **Hardness** 7
> **Density** 3.2
> **Luster** Vitreous, silky in aggregates
> **Cleavage** Perfect, but not recognizable due to fibrous habit
> **Fracture** Uneven
> **Tenacity** Brittle

Similar minerals
Asbestos fibers are flexible; beryl is hexagonal; disthene always shows clear differences in hardness; diopside has a different cleavability.

Orthorhombic crystal form

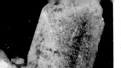

Euclase

AlBe[OH/SiO₄]

Euclase forms colorless, light green or blue crystals, which are prismatic to tabular and usually strongly vertically striated. Euclase primarily only exhibits raised crystals that very rarely are coarse.

Monoclinic crystal form

double-ended euclase crystal

quartz crystal

striations

tabular crystal

Similar minerals
In contrast to the vertically striated euclase, quartz crystals are always horizontally striated; they also have no cleavability; albite has a different crystal form.

Spinel
$MgAl_2O_4$

Spinel mainly forms octahedron crystals and twins, which are mostly intergrown. Placers contain rounded crystals. Spinel varies in color depending on its composition. Iron makes it purple to black and zinc makes it greenish. Pure magnesium spinel is colorless to red and is also called noble spinel.

red spinel octahedron

Occurrences *In metamorphic rocks, in marbles and calc-silicate rocks, in placers.*

> ***Hardness*** *8*
> ***Density*** *3.6*
> ***Luster*** *Vitreous*
> ***Cleavage*** *Hardly recognizable*
> ***Fracture*** *Conchoidal*
> ***Tenacity*** *Brittle*

Isometric crystal form

Similar minerals
Corundum has a different crystal form; cut ruby cannot be distinguished from red spinel by simple methods.

purple spinel octahedron

207

vitreous luster

calcite

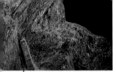

Beryl
$Al_2Be_3[Si_6O_{18}]$

Occurrences In pegmatites, pneumatolytic and hydrothermal veins, in mica schists.

> **Hardness** 7.5–8
> **Density** 2.63–2.8
> **Luster** Vitreous
> **Cleavage** Somewhat recognizable basal cleavage
> **Fracture** Conchoidal to uneven
> **Tenacity** Brittle

Beryl forms hexagonal crystals, which can appear prismatic to tabular and rarely are rich in faces. They are mostly embedded (cloudy), rarely raised (transparent). Beryl can occur in giant crystals up to several meters in size and can weigh several tons. These giant crystals are used to extract the element beryllium, i.e. it is extracted as beryllium ore.

Hexagonal crystal form

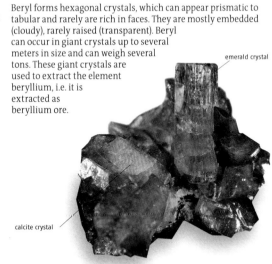

emerald crystal

calcite crystal

rhyolite

red beryl

Beryllium grows in pegmatites (common beryl and aquamarine), in pegmatite druses (morganite, pezzottaite, aquamarine, golden beryl) and in mica slate and hydrothermal calcite veins (emerald). The best European emeralds come from the famous deposit of the Leckbachscharte in the Habach valley in Hohe Tauern, Austria, where the crystals are embedded in mica schist. The emeralds from the deposits in the Ural Mountains are also famous. Beautifully colored and transparent beryls are cut into precious stones.

Occurrences In pegmatites, pneumatolytic and hydrothermal veins, in mica schists.

> *Hardness* 7.5–8
> *Density* 2.63–2.8
> *Luster* Vitreous
> *Cleavage* Somewhat recognizable basal cleavage
> *Fracture* Conchoidal to uneven
> *Tenacity* Brittle

Hexagonal crystal form

209

heliodor

prismatic crystal

aquamarine crystal

prismatic crystal

Similar minerals
Apatite is much softer; quartz hardly forms embedded crystals and is never blue or green; otherwise the formation of six-sided crystals is very distinct. Dioptase always has three-sided end faces and is significantly softer; topaz has excellent cleavage properties and clearly orthorhombic crystals.

Phenakite

Be₂[SiO₄]

Occurrences *In mica schist, in pegmatites and granites, in alpine-type fissures.*

> **Hardness** 8
> **Density** 3.0
> **Luster** *Vitreous*
> **Cleavage** *None*
> **Fracture** *Conchoidal*
> **Tenacity** *Brittle*

The white, colorless, yellowish and, more rarely, pink crystals are prismatic to tabular and lenticular. Its prisms are vertically striated. The crystals are embedded and raised. Phenakite often forms twins in which the crystals are intergrown in a way that the corners of one crystal protrude through the end faces of the other.

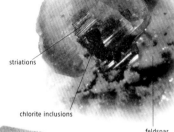

striations

chlorite inclusions

feldspar

Trigonal crystal system

Similar minerals
Quartz is somewhat softer and always striated on the prisms; apatite is softer; beryl is not trigonal, but hexagonal.

210

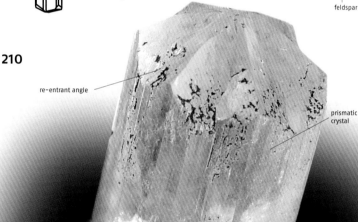

re-entrant angle

prismatic crystal

Chrysoberyl, Alexandrite
Al_2BeO_4

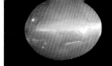

The yellowish to green chrysoberyl crystals are prismatic to thickly tabular. There are heart-shaped to v-shaped twins and triplets resemble hexagonal bipyramids. The crystals are mostly embedded, rarely raised. Alexandrite has the ability to change color: It is green in daylight and changes to red in incandescent light. Chrysoberyl, especially alexandrite, is cut into precious stones.

Occurrences In pegmatites and mica schists.

> *Hardness* 8.5
> *Density* 3.7
> *Luster* Vitreous
> *Cleavage* Recognizable basal cleavage
> *Fracture* Conchoidal
> *Tenacity* Brittle

Similar minerals
The high hardness of chrysoberyl leaves little room for confusion; topaz always has very good cleavability; beryl has a completely different hexagonal crystal form; quartz is softer and has a different crystal form.

alexandrite under incandescent light

mica schist

Orthorhombic crystal form

211

re-entrant angle

chrysoberyl drilling

Topaz

$Al_2[F_2/SiO_4]$

Occurrences *In pegmatites, in pneumatolytic formations, rounded in placers.*

> **Hardness** *8*
> **Density** *3.5–3.6*
> **Luster** *Vitreous*
> **Cleavage** *Perfect basal*
> **Fracture** *Conchoidal*
> **Tenacity** *Brittle*

Orthorhombic crystal form

Topaz crystals are short or long columnar and are raised and embedded. Giant crystals can weigh up to many hundreds of kilograms. More rarely, topaz is radial (pycnite) and coarse. Colors vary from colorless to white, yellow, blue, green, red, pink, purple to brown. Beautifully colored transparent crystals are cut to gemstones.

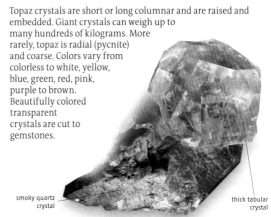

smoky quartz crystal

thick tabular crystal

Similar minerals
Quartz is lighter and has no cleavability; fluorite is much softer; beryl has a completely different crystal form and not so good cleavability.

prismatic crystal

rhyolite

Corundum (Ruby)

Al_2O_3

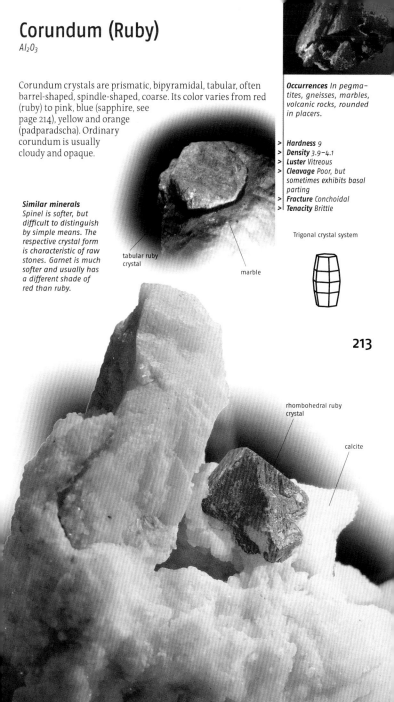

Corundum crystals are prismatic, bipyramidal, tabular, often barrel-shaped, spindle-shaped, coarse. Its color varies from red (ruby) to pink, blue (sapphire, see page 214), yellow and orange (padparadscha). Ordinary corundum is usually cloudy and opaque.

Occurrences In pegmatites, gneisses, marbles, volcanic rocks, rounded in placers.

> **Hardness** 9
> **Density** 3.9–4.1
> **Luster** Vitreous
> **Cleavage** Poor, but sometimes exhibits basal parting
> **Fracture** Conchoidal
> **Tenacity** Brittle

Trigonal crystal system

Similar minerals
Spinel is softer, but difficult to distinguish by simple means. The respective crystal form is characteristic of raw stones. Garnet is much softer and usually has a different shade of red than ruby.

tabular ruby crystal

marble

213

rhombohedral ruby crystal

calcite

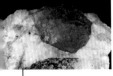

Corundum (Sapphire)

Al_2O_3

Occurrences In pegmatites, gneisses, marbles, volcanic rocks, rounded in placers.

> **Hardness** 9
> **Density** 3.9–4.1
> **Luster** Vitreous
> **Cleavage** Poor, but sometimes exhibits basal parting
> **Fracture** Conchoidal
> **Tenacity** Brittle

Transparent varieties of corundum are cut into precious stones. While only the red variant is called ruby (see page 213), the name sapphire is used for all other colors: white sapphire, yellow sapphire, pink sapphire, etc. The common blue sapphire is simply called sapphire. Some rubies and sapphires are cabochons with a six-rayed star of light. Sapphire is usually fired to produce the beautiful blue color.

Trigonal crystal system

faceted sapphire

tabular sapphire crystal

214

Similar minerals
Sapphire is much harder than any other blue mineral.

feldspar

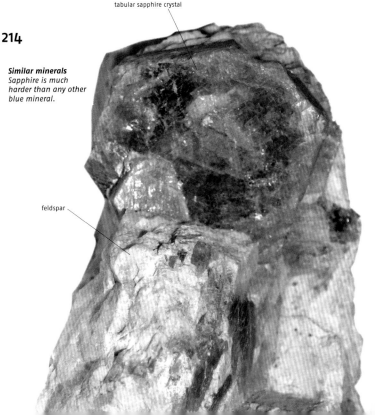

Diamond
C

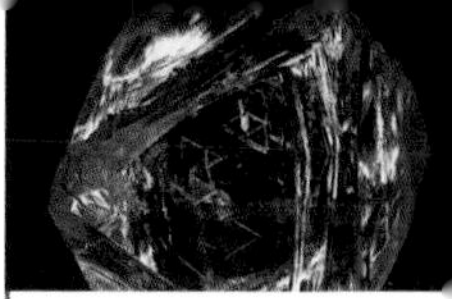

Diamonds almost always occur in more or less well formed crystals. The most common are octahedrons, rhombic dodecahedrons, rarely cubes. It's dinguishable by its strongly etched, rounded crystals and radial aggregates (bort). The crystals are always embedded, never raised. Colors vary from colorless to white, yellow, brown, reddish, greenish, blue, gray to black. Pure diamonds are cut into brilliant-cut diamonds.

Occurrences In kimberlites, in eclogites, in placers, in conglomerates.

> **Hardness** 10
> **Density** 3.52
> **Luster** Adamantine
> **Cleavage** Perfect octahedral
> **Fracture** Conchoidal
> **Tenacity** Brittle

diamond rhombic dodecahedron

Isometric crystal form

Similar minerals
Its high hardness distinguishes diamond from all other minerals.

etched surface

octahedral crystal

kimberlite

Granite
Igneous rock

Occurrences *Smaller and larger intrusions, igneous intrusions, domes, veins.*

Constituent minerals
> ***Primary*** *Potassium feldspar, plagioclase, quartz*
> ***Secondary*** *Muscovite, biotite, hornblende, tourmaline*

Granite is formed through the melting of rocks with a granitic composition as the final stage of metamorphosis. Granite is medium to coarse-grained, often containing foreign rock inclusions, often strongly fissured. It is used as a building stone, for decoration purposes, for gravestones, curbs and as gravel in road construction.

Similar rocks
In granodiorite, plagioclase predominates over potassium feldspar, while gneiss shows clear schistosity.

Colors: white, gray, reddish, greenish, yellowish

fine-grained biotite granite

potassium feldspar (flesh-colored)

plagioclase (white)

quartz (gray)

Porphyritic Granite
Igneous rock

Porphyritic granite is formed by melting rocks with a granitic composition as the final stage of metamorphosis. It has a medium to coarse-grained matrix, porphyric with large potassium feldspars, often containing foreign rock inclusions, and is often strongly fissured. It is used as a building stone, for decoration purposes, for gravestones, curbs and as gravel in road construction.

spheroidal granite

Occurrences *Smaller and larger intrusions, igneous intrusions, domes, veins.*

> **Inclusions** *Potassium feldspar, plagioclase, hornblende, tourmaline*
  Constituent minerals
  Primary *Potassium feldspar, plagioclase, quartz*
> **Secondary** *Muscovite, biotite, hornblende, tourmaline*

Colors: white, gray, reddish, greenish, yellowish

Similar rocks
In granodiorite, plagioclase predominates over potassium feldspar, while gneiss shows clear schistosity.

biotite (black)

plagioclase (white)

quartz (gray)

potassium feldspar (reddish)

Aplite
Igneous rock

Occurrences As veins in granites and their surrounding rocks, common in all granite sites.

> **Inclusions** Tourmaline, hornblende
> **Constituent minerals**
> **Primary** Quartz, potassium feldspar
> **Secondary** Biotite, muscovite, hornblende, tourmaline

Colors: white to light gray

When a magma body slowly solidifies, remains of the still liquid magma penetrate into cracks and crevices in the surrounding rocks. This forms dike rocks, such as the bright aplite. It is fine-grained, sometimes zoned, dike-shaped, often also as a fine-grained zone next to or around pegmatite.

garnet (reddish brown)

aplite dike

granite

Similar rocks
Aplite's characteristic dike-shaped appearance, together with its light color, makes it unmistakable. Pegmatite is much more coarse grained.

Pegmatite
Igneous rock

Occurrences As veins, streaks, igneous intrusions, mostly in combination with granites.

> **Inclusions** Tourmaline, columbite, beryl, topaz, lepidolite and many other rare minerals
> **Constituent minerals**
> **Primary** Quartz, potassium feldspar
> **Secondary** Plagioclase, muscovite

Colors: white, gray, pink, many different colors

Volatile phases remain at the end of rock crystallization. These include all the elements that cannot be used to form normal rock-forming minerals. From these, coarse to giant pegmatites form in crevices, cracks and other cavities. Granite with feldspar and quartz intergrowths that resemble script is called graphic granite.

graphic granite

potassium feldspar

quartz

Similar rocks
The large granularity of pegmatite does not allow any confusion.

Syenite
Igneous rock

Syenite is formed from more alkaline magmas. It is medium to coarse-grained, rarely porphyric, sometimes druse, porous. It is used locally as a building material and for gravel production. Nepheline is the main component in nepheline syenite.

nepheline syenite

aegirine (black)

nepheline (white)

Similar rocks
In contrast to syenite, granite's primary component is quartz, while hornblende is at best a secondary component. Diorite, unlike syenite, does not contain potassium feldspar as its primary component.

> **Occurrences** *In smaller separate intrusive bodies, as part of large differentiated gabbro rock bodies.*

> **Inclusions** *Hornblende, pyroxene, titanite*
> **Constituent minerals**
> **Primary** *Potassium feldspar, plagioclase (andesine–oligoclase), hornblende*
> **Secondary** *Biotite, pyroxene, quartz*

Colors: light to dark gray

feldspar (white)

hornblende (black)

hornblende syenite

Diorite
Igneous rock

Diorite is the first material to be excreted from granitic magmas. It is fine to medium grained, rarely shows a spherical structure (spherical diorite). Locally it is used as a building material, for the production of gravel — beautifully colored variants are also used as a decorative stone.

titanite (brown with white halo)

fine-grained matrix

Colors: medium to dark gray

Similar rocks
Gabbro contains anor-thite-rich plagioclase and pyroxene as its primary constituent in contrast to hornblende in diorite.

diorite with spots of titaniumite

feldspar (white)   hornblende (black)

medium-grained diorite

Gabbro
Igneous rock

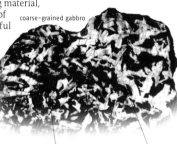

Gabbro is formed from ultrabasic magmas in the earth's mantle. It is medium to coarse-grained, and the feldspars are often distributed in the rock-like strips, sometimes porphyric, often banded, and Gabbro often shows flow structures. It is used locally as a building material, for the production of gravel. More beautiful variants are also used as a decorative stone in architecture and for gravestones.

coarse-grained gabbro

pyroxene (black)

plagioclase (white)

Similar rocks
Diorite contains horn-blende as the primary constituent instead of pyroxene; norite contains orthorhombic pyroxene; pyroxenite contains no feldspar.

> **Occurrences** *In large stratified basic intrusions, as independent rock bodies.*

> **Inclusions** *Plagioclase, pyroxene*
Constituent minerals
> **Primary** *Plagioclase (labradorite-bytownite), pyroxene (monoclinic)*
> **Secondary** *Hornblende, magnetite, ilmenite*

Colors: gray, green, black–white speckled, black

221

fine-grained gabbro

Anorthosite
Igneous rock

Anorthosite is produced from alkaline magmas. It is medium to coarse-grained, always uniform, and interspersed, albeit rarely, with magnetite or chromite strata. If there is sufficient chromite content, it is extracted for chromium production. Beautifully colored and patterned varieties serve as decorative stones in architecture and are used to make gravestones.

Similar rocks
Granite and aplite always contain quartz and potassium feldspar instead of plagioclase; pegmatite is always coarse to very coarse-grained and consists of potassium feldspar.

Colors: white, gray, black, greenish, reddish

222

shiny cleavage surfaces of plagioclase

plagioclase (black)

plagioclase (white)

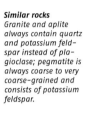

pyroxene (black)

Peridotite

Igneous rock

Peridotite is formed by alkaline magmas transported from the upper mantle of the earth. Dunite is a related rock that consists only of olivine. It is assumed that the earth's mantle is partly composed of peridotite rocks. The olivine bombs we find in volcanic rocks are fragments of such rocks. Peridotite is sometimes used as a decorative stone.

fine-grained olivine
(olive green)

dunite

Occurrences As smaller independent rock complexes, in ophiolite complexes, in basic rock complexes.

> *Inclusions* Pyrope, pyroxene
> *Constituent minerals*
> *Primary* Olivine, pyroxene
> *Secondary* Spinel, hornblende, pyrope, phlogopite, chromite

Colors: light to dark gray

pyrope (red)

fine-grained olivine
(gray green)

Similar rocks
Gabbro contains feldspar.

Gneiss
Metamorphic rock

Gneiss is medium to coarse-grained, with light and dark foliation, as well as streaky, folded. It shows some sprinklings of feldspar (augen gneiss) or garnet. Gneiss is formed from clayey sediments (paragneiss) or granitic rocks (orthogneiss) during medium to high-grade metamorphosis. Well-foliated, unfolded gneisses are used as floor and roof slabs.

biotite (black)

garnet (red)

Similar rocks
Granite is not foliated.

folded gneiss

feldspar (white)

biotite (black)

Cordierite Gneiss
Metamorphic rock

During high-grade metamorphosis, cordierite gneiss is formed from clay sediments. It is medium to coarse-grained and often displays light and dark foliation. Cordierite gneiss is often streaked, intensely folded and has more or less blue-colored inclusions of cordierite, which can sometimes reach several centimeters in size.

fractures perpendicular to the stratification

biotite (black)

feldspar (yellowish white)

Occurrences *In all highly metamorphic areas.*

> ***Inclusions*** *Cordierite, garnet*
> ***Constituent minerals***
> ***Primary*** *Feldspar, quartz, mica*
> ***Secondary*** *Garnet, cordierite, sillimanite, hornblende*

Colors: light to dark gray, brownish, many different colors

feldspar (yellowish white)

biotite (black)

cordierite

fracture parallel to stratification

225

Similar rocks
Granite is not foliated, garnet amphibolite consists mainly of amphiboles, garnet gneiss is better arranged.

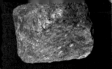

Mica Schist
Metamorphic rock

Occurrences *Frequently in regional metamorphic areas, e.g. in the Alps.*

> **Inclusions** *Garnet, disthene, hornblende*
Constituent minerals
> **Primary** *Mica, quartz*
> **Secondary** *Feldspar, chlorite, garnet, tourmaline, actinolite, hornblende, disthene*

Colors: gray, silver gray, black, brown, glossy

Mica schist is formed during medium to high grade metamorphosis from sandy to clayey parent rocks. It is fine- to coarse-grained, often folded, sometimes with layers rich in quartz or feldspar. It often contains large inclusions of garnet, hornblende or disthene.

Similar rocks
Phyllite, in contrast to mica schist, does not show the individual mica flakes with a magnifying glass, nor does phyllite have any inclusions. Gneisses always contain feldspar as the primary constituent.

hornblende (black)

mica (white)

hornblende mica schist

garnet (reddish brown)

mica (silvery)

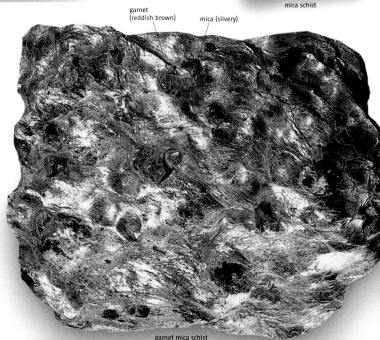

garnet mica schist

Phyllite
Metamorphic rock

Phyllite is formed during low-grade regional metamorphism from clayey to sandy sedimentary rocks. It is very fine-grained. The individual mica flakes are not visible even with a magnifying glass. It is slaty, folded in layers, and often very finely corrugated. Finely crushed, it is used for coating highly reflective cardboard and mats.

Occurrences In areas with extensive metamorphosis (regional metamorphosis).

Constituent minerals
> **Primary** Quartz, mica
> **Secondary** Graphite, feldspar, chlorite, chloritoid

Similar rocks
Slate doesn't have a silky luster like phyllite; in mica schist you can distinguish the individual mica flakes with a magnifying glass; in contrast to phyllite, it also often has inclusions of various minerals.

folds

heavily folded phyllite

Colors: gray, yellowish, greenish, silvery, often silky luster

silky shimmer

cleavage plane

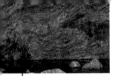

Amphibolite
Metamorphic rock

Constituent minerals
> **Primary** Amphibole (hornblende, actinolite)
> **Secondary** Epidote, plagioclase, chlorite, garnet

Amphibolite is formed during low to moderate metamorphosis from basic, mostly volcanic rocks. It is coarse-grained, foliated, sometimes interspersed with garnet inclusions. It is used, albeit rarely, for gravel production and as a building material.

plagioclase (white)

amphibole (gray-green)

Colors: dark green to black

fine-grained amphibolite

Similar rocks
Serpentinites do not contain amphiboles; chlorite schist has chlorite as its primary constituent and is much softer than amphibolite; eclogite has pyroxene as its primary constituent and does not contain amphiboles.

228

garnet (red)

hornblende (black)

plagioclase dike filling (white)

Eclogite
Metamorphic rock

Eclogite is formed from basic rocks during high-grade metamorphism. It is coarse-grained; exhibits garnet inclusions, more rarely of disthene inclusions and is rarely stratified. Minerals of lower temperatures and pressures, such as mica or disthene, are often formed during cooling. Eclogite is often used as decorative stone.

Occurrences *Lenses and layers within highly metamorphic rock sequences and bodies.*

> ***Inclusions*** Garnet, disthene
> ***Constituent minerals***
> ***Primary*** Pyroxene (omphacite), garnet
> ***Secondary*** Disthene, quartz, actinolite

Colors: light to dark gray, red speckled

garnet inclusions (brown)

rounded eclogite pebble

229

garnet inclusions (red)

omphacite (green)

unlayered eclogite

Similar rocks
The characteristic composition does not allow for any confusion.

Marble
Metamorphic rock

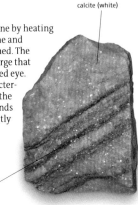

calcite (white)

graphite layer
(black)

Occurrences *In the contact aureole around deep rocks, in regional metamorphic rock formations.*

> ***Inclusions*** *Spinel, garnet, wollastonite*
> **Constituent minerals**
> ***Primary*** *Calcite*
> ***Secondary*** *Dolomite, wollastonite, vesuvianite, graphite, diopside, spinel, corundum*

Marble is formed from limestone by heating during metamorphosis. It is fine and coarse-grained, sometimes zoned. The individual crystallites are so large that you can see them with the naked eye. This gives the marble its character-istic sparkle. Marble is used in the building trade, alongside all kinds of limestones that are incorrectly referred to as marble.

Colors: white, yellowish, brownish

Similar rocks
In contrast to marble, limestone does not show the cleavage surfaces of the indi-vidual calcite grains; gypsum rock is softer.

calcite

shiny cleavage surfaces of calcite

Silicate marble (calc-silicate rock) contains larger amounts of silicate mixtures (e.g. olivine, diopside, wollastonite, grossular and vesuvian); dolomite marble contains dolomite as the primary constituent; tremolite is often present as an inclusion; corundum is rare. Marble is used as a building material, for decoration purposes, for gravestones and sculptures.

Occurrences In the contact aureole around deep rocks, in regional metamorphic rock formations.

> *Inclusions* Spinel, garnet, wollastonite
> **Constituent minerals**
> **Primary** Calcite
> **Secondary** Dolomite, wollastonite, vesuvianite, graphite, diopside, spinel, corundum

Colors: white, yellowish, brownish

231

fine-grained

dolomite marble

wollastonite (white)

grossular (brown)

diopside (gray-green)

silicate marble

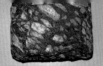

Chlorite Schist

Metamorphic rock

Occurrences In areas of extensive metamorphosis.

> **Inclusions** Magnetite, pyrite
> **Constituent minerals**
> **Primary** Chlorite
> **Secondary** Magnetite, pyrite, hornblende, epidote, albite

Chlorite schist is formed during low-grade metamorphosis of lavas, volcanic tuffs and other basic rocks. It is fine to coarse-grained, flaky, slaty, often with magnetite and pyrite inclusions.

Colors: gray-green to dark green

Similar rocks
Mica schist and phyllite have mica as their primary mineral, amphibolites contain hornblende or actinolite as their primary constituents. Its green color and the low hardness (hardness = 2) make chlorite schist unmistakable.

pyrite (golden yellow)

coarse-flaked chlorite schist

magnetite (black)

chlorite (gray-green)

Sandstone, Sand

Sedimentary rock

Sandstone is fine- to medium-grained, layered and bedded. The grains of sand can be cemented with quartz (quartz sandstone) or calcite (limestone sandstone) or clay (clay sandstone). Sandstones with feldspar particles are called arkose. Sandstones are formed by solidifying sand, which is considered to be loose rock. Sandstone is used in a variety of ways as a building material: as floor slabs, for sculptural work. Traces of living organisms, so-called trace fossils, are common in sandstone.

Occurrences *In all sedimentary strata, always formed within continental regions.*

> **Inclusions** *Feldspar, mica*
> **Constituent minerals**
> **Primary** *Quartz grains*
> **Secondary** *Mica, feldspar, calcite*

— feldspar (white)

arkose

Colors: white, yellow, red, brown, green

Similar rocks
Breccias and conglomerates consist of rock fragments.

colored red by iron oxides

red sandstone

Gravel, Conglomerate
Sedimentary rock

Occurrences *In younger sedimentary layer sequences.*

Constituent minerals
> **Primary** *Pebbles, up to several centimeters in grain size*
> **Secondary** *Sand, clay minerals*

Gravel is granular and layered. Without a binding agent, it is loose rock. It is formed during the accumulation of sediments caused by the erosion of silicate rocks and limestones, often as a result of glaciation. Gravel consolidated by cement is called conglomerate. Gravel is used in many ways: for construction purposes, in road construction and as an additive in concrete production.

Colors: white, multicolored

rounded rock fragments (pebbles)

limy cement

conglomerate

Similar rocks
Sandstones are solid, sand has much smaller grains. Conglomerates and nagelfluh are solidified gravel.

pebbles

well rounded

Breccia
Sedimentary rock

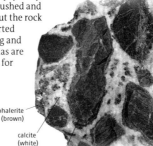

Breccias are coarse-grained, with varying grain sizes, and all rock fragments are angular, the binding matrix can be sandy, chalky or clayey. They are created when rocks are crushed and then reconsolidate without the rock fragments being transported away. Mechanically strong and optically attractive breccias are used as decorative stones for wall coverings.

Occurrences In areas with high mechanical stress.

> *Inclusions* Rock fragments
Constituent minerals
> *Primary* Rock fragments
> *Secondary* Calcite, quartz

sphalerite (brown)
calcite (white)
limestone fragment (black)

Colors: various colors, also very colorful

Similar rocks
Conglomerates consist of rounded rock fragments.

angular limestone fragments (gray)

limy cement

Slate, Shale
Metamorphic rock/Sedimentary rock

Shale is formed by the deposition of clay minerals in water, especially in the sea. It contains numerous flattened fossils. When the shale is heated due to an increase in pressure at a given depth, slate forms; which is even denser, hardly contains any fossils and is easily cleavable. Slate slabs are used to cover roofs, weatherproof cladding for houses, and is used for table tops and floor tiles.

bedding plane

slate

slate with petrified fish

236

conchoidal fracture

stratification

shale

Similar rocks
Phyllites display a rich, silvery shimmer of mica on the layered surfaces.

Dolomite
Sedimentary rock

Dolomite is primarily formed from limestone via magnesium exchange with magnesium-containing water or rock. It is rarely formed as dolomite directly. It is fine to medium-grained, layered and rarely folded. It serves as a building block, for the production of gravel, as floor tiles, for the production of dolomite bricks for blast furnaces and as a so-called aggregate in steel smelting.

wave dolomite

fine-grained dolomite

Limestone
Sedimentary rock

Colors: white, yellowish, brownish, gray, black

Limestone is fine-grained, layered, and sometimes folded. Rarely it is completely dense and unstructured. It usually is made up of the remains of living organisms; rarely is it inorganic. Limestone is used as building material, as gravel and for lime burning. Beautifully colored and patterned varieties also are used as ornamental stone, for wall coverings and as floor tiles.

fossilized ammonites

ammonite limestone

238

Similar rocks *In contrast to limestone, dolomite does not effervesce when dabbed with diluted hydrochloric acid.*

colored red by iron oxides

fossilized coral

coral limestone

Limestone often contains numerous remains of living creatures that can be seen with the naked eye, therefore, it is also often named snail-shell limestone, ammonite limestone or coral limestone. Limestones with platy cleavage are also referred to as platy limestone. Calcareous deposits by hot springs, which often still contain remains of plants, are called calcareous tufa.

Occurrences *In many sedimentary strati-graphic sequences.*

> ***Inclusions*** *Pyrite, flint, marcasite, quartz, fossils*
> ***Constituent minerals***
> ***Primary*** *Calcite*
> ***Secondary*** *Limonite, dolomite, quartz, clay minerals*

Colors: white, yellowish, brownish, gray, black

239

calcareous oolite (spherical)

cement matrix colored brown by iron oxides

colored yellowish by iron

conchoidal fracture

dense fine-grained limestone

Rhyolite
Volcanic rock

Occurrences In vents, intrusions, veins, rarely forming regular rock formations.

> **Inclusions** Potassium feldspar
Constituent minerals
> **Primary** Quartz, potassium feldspar
> **Secondary** Plagioclase (albite), biotite

Colors: very light gray to whitish, light brown

Rhyolite's matrix is very fine-grained and sometimes shows large inclusions of sanidine (potassium feldspar). Rhyolite is used locally as a building material and for gravel production.

fine-grained matrix

sanidine (white)

rhyolite, striped layering

Similar rocks
Rhyolite's matrix is always finer-grained than granite, it never occurs in the area of volcanic activity.

Quartz Porphyry
Volcanic rock

Occurrences As gigantic lava plateaus, especially from the Permian and Triassic periods about 200 million years ago.

> **Inclusions** Quartz, potassium feldspar
Constituent minerals
> **Primary** Quartz, potassium feldspar
> **Secondary** Plagioclase (albite), biotite

Colors: brown, reddish brown, the matrix is often colored by iron oxides

Similar rocks
Rhyolite doesn't have a reddish matrix; quartz porphyry has a finer-grained matrix than granite.

Quartz porphyry has a fine-grained matrix with inclusions of quartz and potassium feldspar. It is formed when silica-rich magmas rise, which are able to cover large areas due to their high mobility. Quartz porphyry is the name for geologically ancient rhyolites. It is used locally as a building material, for the production of paving stones, floor tiles and gravel.

feldspar (reddish

colored red by iron

quartz (gray)

fine-grained matrix

Phonolite
Volcanic rock

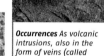

Phonolite is fine-grained with individual inclusions of nepheline and potassium feldspar. It shows typical conchoidal fracture, often flow structures, and sometimes columnar parting. It is formed from alkaline-rich magmas. Phonolite is the volcanic rock that corresponds to the plutonic rock nepheline syenite. It is used as a building material and for gravel extraction.

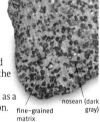

nosean (dark gray)

fine-grained matrix

fine-grained matrix

nepheline (dark gray)

Similar rocks
In contrast to phonolite, tephrite contains leucite, often as large inclusions.

Occurrences *As volcanic intrusions, also in the form of veins (called tinguaite).*

> ***Inclusions*** Hauyne, potassium feldspar, nepheline, melanite (a black garnet containing titanium)
Constituent minerals
> **Primary** Nepheline, potassium feldspar, aegirine (a sodium pyroxene)
> **Secondary** Olivine, sodalite, hauyne, sodium hornblende

Colors: light to dark gray, greenish, brown

241

Basalt
Volcanic rock

Basalt is formed when gabbro-like magmas rise. Basalt is the volcanic rock that corresponds to the plutonic rock gabbro. It is dense with a conchoidal fracture, sometimes scoriaceous with a rough surface, its matrix is very fine-grained, often showing columnar parting.

fine-grained basalt

dense olivine basalt

olivine (yellow-green)

Similar rocks
All deep rocks, which are similar in principle, are much coarser-grained; rhyolite contains quartz as its primary constituent, tephrite and leucite.

Occurrences *In the form of lava flows, nappes, intrusions, veins.*

> ***Inclusions*** Quartz, potassium feldspar
Constituent minerals
> **Primary** Quartz, potassium feldspar
> **Secondary** Plagioclase (albite), biotite

Colors: brown, reddish brown, the matrix is colored by iron oxides

Andesite
Volcanic rock

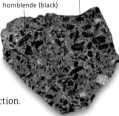

fine-grained matrix

hornblende (black)

Occurrences *In rivers, effusions, laccolith.*

> **Inclusions** *Plagioclase, pyroxene, hornblende*
Constituent minerals
> **Primary** *Plagioclase, pyroxene, hornblende*
> **Secondary** *Biotite, magnetite*

Andesite's matrix is very fine-grained, at times glassy, sometimes with large inclusions of hornblende, plagioclase and pyroxene. Andesite is formed during the melting of oceanic crust in subduction zones. It is used locally as a building material and for gravel production.

hornblende andesite

Colors: brown to brown–black

fine-grained andesite

Similar rocks
Dacite and rhyolite contain quartz, latite contains sanidine.

242

Lava
Volcanic rock

Occurrences *Area of young, partly still active volcanoes.*

> **Inclusions** *Olivine, augite, hornblende, biotite*
Constituent minerals
> **Primary** *Plagioclase, pyroxene*
> **Secondary** *Olivine, hornblende, biotite*

Lava refers to volcanic rocks that have solidified on the surface, more or less basaltic in composition. It shows jagged, flat-like, rope-like solidification forms with diverse flow structures. It is dense to porous, fine-grained and can contain various inclusions.

jagged surface

loose inner mass

Colors: black, gray, brown

Similar rocks
The surface structure makes lava stone unmistakable.

dense outer layer

pahoehoe lava

Obsidian
Volcanic rock

Obsidian is formed during very rapid cooling of magmas rich in silicic acid. It is dense with a conchoidal fracture, sometimes scoriaceous with a rough surface, translucent to opaque. Beautiful varieties can also be used as gemstones or cut into art objects. Varieties have many names, depending on color and appearance: Mahogany obsidian, snowflake obsidian, rainbow obsidian, etc.

Occurrences *In lava flows, lava crusts, ejecta.*

> **Inclusions** *Cristobalite*
> **Constituent minerals**
> **Primary** *Rock glass*
> **Secondary** *Cristobalite, magnetite*

feldspar

obsidian

snowflake obsidian cabochon

Colors: black, brown, white speckled

rainbow obsidian

sharp edge

conchoidal fracture

Similar rocks
The glassy texture makes confusion impossible.

Pumice
Volcanic rock

Occurrences In layers
in the ejecta masses of
silica-rich volcanoes.

> **Inclusions** Hauyne,
titanite
Constituent minerals
> **Primary** Rock glass
> **Secondary** Sanidine,
hornblende, pyroxene

Colors: white, gray

Pumice stone is very porous, rough, extremely light, and even
floats on water. It is formed during volcanic eruptions with a high
gas content. It can be used in many ways: for the production of
lightweight building materials, as an additive to building materials,
and as a planting material.

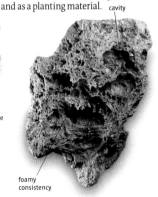

cavity

cavity

foamy
consistency

close-up view of foamy pumice stone

Similar rocks
*The glassy, highly porous
nature makes mix-ups
impossible; no other rock
floats on water.*

Volcanic Tuff
Volcanic rock

Occurrences In the
vicinity of volcanoes.

Constituent minerals
> **Primary** Ejecta, volcanic
ash, rock glass
> **Secondary** Augite, horn-
blende, sanidine, olivine

Colors: black to gray-black,
brown, gray

Volcanic tuff can have different grain sizes and can be porous,
often nicely stratified. It is formed from the deposition and
solidification of loose masses ejected by volcanoes. If these
masses are still molten, then so-called welded tuff or ignimbrite
is formed. Sometimes volcanic tuffs are used as building
materials, for cement production and for concrete.

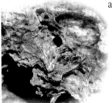

ignimbrite
(welded
tuff)

Similar rocks
*Calcareous tufa effer-
vesces when dabbed
with diluted hydro-
chloric acid.*

brown
volcanic tuff

Coal
Sediment

Coal is formed from plant matter under the absence of air, so-called coalification. Peat is produced first, then lignite, then hard coal. Hard coal is usually geologically older than lignite. When the temperature rises, e.g. during mountain formation processes, geologically younger lignite can also be converted into hard coal. This is called pitch coal. Dense coal is also polished as a gemstone, known as jet.

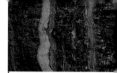

pitchy luster

wood texture

polished jet

lignite

> **Occurrences** In numerous sedimentary sequences, hard coal, especially in carbon.

> **Inclusions** Pyrite
> **Constituent minerals**
> **Primary** Carbonization of organic matter, especially parts of plants
> **Secondary** Quartz, calcite, pyrite

Colors: brown, black

Serpentinite
Metamorphic rock

Serpentinite is fine to coarse-grained, fibrous, flaky, dense and very tough. It is formed by low-grade metamorphosis from peridotites. Beautifully dyed or patterned serpentinites are cut into gemstones or handcrafted items.

serpentine (yellow-green)

dense antigorite

calcite (white)

Similar rocks
Amphibolites consist of amphiboles as their primary constituents.

> **Occurrences** In regional metamorphic areas.

> **Inclusions** Olivine
> **Constituent minerals**
> **Primary** Serpentine as antigorite, rarer than chrysotile
> **Secondary** Magnetite, chromite, olivine, talc, dolomite, magnesite

Colors: gray-green, yellow-green, dark green, brown

Accompanying minerals Minerals that frequently occur in combination with the mineral described

Aggregate Grouping of several crystals. It can be, for example, radial, fibrous, spherical or reniform

Alpine-type fissure *See* fissure

Amorphous Minerals that do not have a crystal structure, such as opal, are called amorphous. They are not crystalline

Amphibole Group of silicate minerals with a typical cleavage angle of about 120°

Cementation zone Underground area of a deposit where certain elements are concentrated

Cleavage *See* Introduction

Coarse Mineral pieces, which are not limited by crystal surfaces but by irregular fracture surfaces, are coarse

Concretion A rounded, more or less spherical, mineral accumulation in a sedimentary rock

Constituent mineral Minerals from which a rock is composed

Cruciform twinning Twinning, in which the two individual crystals have grown together in a cruciform (cross-shaped) way

Density (specific weight) The weight of a cube of a specific mineral with an edge length of 1 cm

Deposit Accumulation of minerals at a specific point on our planet

Doubly terminated Crystal with a complete termination on both ends

Druse A more or less round cavity in a rock in which crystals grow

Ductile Stretchable, deformable, such as gold

Exhalation Leakage of gases from the earth's interior

Fissure Cavities or cracks in a rock caused by stress conditions. Fissures in silicate rocks, especially in the Alps, are called alpine-type fissures

Fracture Shape of the fracture surface, e.g. conchoidal, uneven, splintery, hackly, granular

Gangue Minerals that accompany ore minerals in an ore vein, e.g. quartz, calcite, barite

Geode A spherical cavity in volcanic rocks lined with crystals

Hardness *See* Introduction

Hydrothermal formations Minerals formed from warm to hot aqueous solutions

Impregnation The process of filling small cavities within a rock with minerals at a later period of time (e.g. an ore)

Inclusion Crystal or fossil embedded in a finer-grained rock

Luster *See* Introduction

Metamorphic Rocks that have been transformed by changes in pressure and/or temperature are called metamorphic rocks

Metamorphism Alteration of rocks or minerals due to pressure and/or heat

Mixed crystals Crystals containing two or more elements in variable proportions, whereby one element replaces the other

Native A metal that occurs in its elemental form in nature, for example gold, silver, copper, platinum

Glossary

Ore deposits Deposits that contain an ore in quantities suitable for mining

Ore Mineral or mixture of minerals used for the extraction of metals or other elements

Oxidation zone Area of a deposit that is exposed to the influence of weathering

Paragenesis Typical occurrence of minerals due to the conditions under which they are formed

Placer Accumulation of heavy, weather-resistant minerals. There are river, sea, beach or surf placers, depending on the location and mechanism of the accumulation. Typical minerals accumulated in placers are gold, diamond, corundum, magnetite, ilmenite, monazite, garnet and chromite

Pneumatolytic formations Minerals that have been formed through the gas phase

Pseudomorph A mineral which chemically replaces another and which has assumed the crystal form thereof

Pyroxene Group of silicate minerals with a typical cleavage angle of about 90°

Radioactive A mineral is radioactive when it emits alpha, beta or gamma rays

Secondary minerals Minerals that have been altered to a new form at the expense of other minerals

Specimen Adhesion of several crystals of the same or different mineral types

Subvolcanic formation Mineral formations that have formed directly below a volcano in the earth's crust

Tarnish Colors that are created by the formation of very thin oxidation skins on minerals, especially sulfides

Tenacity *See* Introduction

Twins Natural adhesion of two crystals of the same mineral type. There are also triplets, quadruplets, quintuplets, etc.

Varieties Minerals that are the same but differ in their special characteristics (color, aggregate form, etc.)

Vein A filling of minerals in a rock void that is younger than the surrounding rock

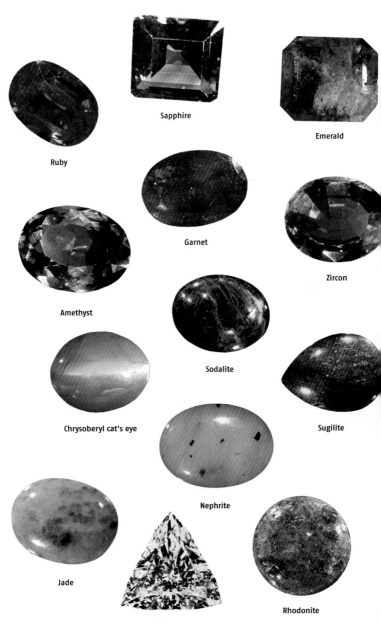

Sapphire

Emerald

Ruby

Garnet

Zircon

Amethyst

Sodalite

Chrysoberyl cat's eye

Sugilite

Nephrite

Jade

Rhodonite

Diamond

Topaz

Aquamarine

Peridot

Opal

Tourmaline

Lapis lazuli

Landscape agate

Jasper

Chalcedony

Turquoise

Malachite

Rainbow obsidian

About the Author

Rupert Hochleitner holds a doctorate in mineralogy. His area of specialization is systematic mineralogy. Other fields of research are meteorites, especially those from the planet Mars, oxidation minerals and pegmatitic phosphate minerals. He has published numerous scientific articles on these subjects. His books on mineral identification have been published in fourteen languages. For many years, Rupert Hochleitner was editor-in-chief of the magazine LAPIS, a trade journal for mineral collectors and mineral lovers. Since 1993, he has been Deputy Director of the Mineralogical State Collection in Munich.

Credits

M = Main graphic, C = Corner graphic, A = Additional graphic, t = Top, b = Bottom

1043 color photos by Rupert Hochleitner, with 43 color photos by Christian Rewitzer on pages: 21 At, 32 Mt, 33 At, 38 Ct, 38 Cb, 38 Mb, 39 A, 40 A, 41 Ct, 41 At, 45 Ab, 46 Ab, 56 Cb, 56 Ab, 57 At, 60 Ct, 60 At, 67 Ab, 69 A, 94 C, 104 A, 107 A, 119 A, 120 A, 121 Ct, 121 At, 121 Cb, 123 Mt,123 Ct, 123 At, 133 At, 133 Ab, 136 Ab, 142 At, 142 Mb, 142 Ab, 161 Mb, 161 Ab, 161 Cb, 165 Ab, 165 Cb, 166 Ab, 166 Cb. 257 crystal drawings by Rupert Hochleitner.

Cover design by Noor Majeed. The main image is andradite. The smaller images, from the top, are dioptase, realgar, and vesuvianite.

The photos on the following pages show: page 1, Vivianite; page 16/17, combinerite; and page 246, plagioclase (labradorite).